WORLD WAR II GERMAN FIELD WEAPONS AND EQUIPMENT

A VISUAL REFERENCE GUIDE

Keith Ward

Helion & Company Ltd

Helion & Company Limited
26 Willow Road
Solihull
West Midlands
B91 1UE
England
Tel. 0121 705 3393
Fax 0121 711 4075
Email: info@helion.co.uk
Website: www.helion.co.uk
Twitter @helionbooks
Blog http://blog.helion.co.uk/

Published by Helion & Company 2014

Designed and typeset by Farr out Publications, Wokingham, Berkshire
Cover designed by Farr out Publications, Wokingham, Berkshire
Printed by Henry Ling Limited, Dorchester, Dorset

Text and images © Keith Ward 2014

ISBN 978-1-909384-44-6

British Library Cataloguing-in-Publication Data.
A catalogue record for this book is available from the British Library.

All rights reserved. No part of this publication may be reproduced, stored in a retrieval system, or transmitted, in any form, and by any means, electronic, mechanical, photocopying, recording or otherwise, without the express written consent of Helion & Company Limited.

For details of other military history titles published by Helion & Company Limited please contact the above address, or visit our website www.helion.co.uk.

We always welcome receiving book proposals from prospective authors.

Contents

Part I – Small Arms
    Mauser Karabiner KAR 98 – standard issue rifle   5
    MP 40 Maschinenpistole   6
    Wimmersperg Spz-kr 7.92mm Assault Rifle   7
    StG 44 (MP 44) (Sturmgewehr 44)   8
    People's Weapons 1945   9
    People's Weapons 1945 – Volkssturmgewehr 1-5   10

Part II - Personal Equipment
    Army Flare Gun   11
    Model 1929 Army Flare Gun (with Sturmpistole Attachment)   12
    1945 Volkssturm Flare Gun   13
    Telephone Wire Spool   14
    Alarmleuchtzeichen   15
    Knijpkat, Dynamo Light   16
    Field lights/torches   17
    Dugout/Bunker - Wall or table light   18
    Field Torch   19

Part III – Grenades
    Defensive and Offensive   20
    Model 34 Stielhandgranate   21
    Model 39 Eihandgranate 'egg' grenade   22
    Model 42 Nelbel Eihandgranate   23
    Rifle Grenades   24
    Blendkörper (blinding device) (smoke) BK-2H/24   25
    Handgranate 343d   26
    Late Production Munitions 1945 - Glass grenade   27
    Improvised Concrete Grenade 1945   28

Part IV - Hand-Held Anti-Tank Weapons
    Panzerbüchse 39 (PzB 39) 1939   29
    Granatbüchse Model 39 (GrB 39) 1942   30
    Panzerfaust 30 (Klein)   31
    Panzerfaust 60   32
    Panzerfaust Family (30, 60, and 150)   33
    Panzerfaust 150   34
    Panzerwurfmine (L)   35
    8.8cm Raketenpanzerbüchse 54/1 (Panzerschreck)   36

Part V – Machine-Guns
    Maschinengewehr 34 (MG 34)   37
    Maschinengewehr 42 (MG 42)   39

Part VI – Mortars
    5cm 'Leichte' Granatwerfer 36   41
    8cm Schwere Granatwerfer 34   42

Part VII - Mines And Demolition Charges
    Improvised Anti-personnel Mines   43
    'Plank' mine and Druckschiene for Teller 1 Iandmine   44
    Holzmine 42   46
    S Mine   47
    Brettstueck mine   48
    Transport Rack for M-6 Teller Mine 42   49
    Teller Mine 42   50
    Glass Mine   51
    Panzerhohlladungsmine   53
    Behelfs-Schützenmine S. 150   54
    Panzerstabmine 43 (Pz.Stab.Mi.43)   55
    Shu-mine 1944   56
    Engineers' tools for demolition, 1944   57
    Topf Mine   58
    Topf Mine Porcelain 1945   59
    Wooden Topf Mine 1945   60
    Hohl-Sprung mine 4672 - hollow-charge bounding mine, 1945   61
    Panzermine A1 - Aluminium Type   62
    Panzermine Steingüt (Pz.Mi Steingüt) Clay Type   63
    Concrete Block Mine   64
    Riegelmine 43 (R.Mi.43), Sprengriegel 43 (Spr.R.43)   65
    Main fuses/Igniters   66
    Panzerschnelmine A and B   68
    Behelfmine E.5   69
    1kg and 3kg Demolition Charges   70
    Flaschenmine 42   71
    Improvised Concrete Stake Mine   72
    Kippzünder 43 Tilt Switch (K.i.43)   73
    Teller mine Booby-trap   74
    Entlastungszünder 44   75
    Haft-Hohlladung 3kg   76
    Stielhandgranate 24, 39, 43   77
    Panzerhandmine 'Sticky mine'   78
    Panzerhandmine 3 Magnetic Mine   78
    Improvised Anti-tank Variants   79

Part VIII - Artillery and crewed anti-tank weapons

3.7cm Pak 35/36 (Panzerabwehrkanone 36)	80
5cm Pak 38 L/60 (1940)	82
7.5cm Pak 40 (7.5cm Panzerabwehrkanone 40)	84
8.8cm PAK 43 (Panzerabwehrkanone 43)	86
Rheinmetall 12.8cm K44 L55	88
2.8cm Schwere Panzerbüchse 41 (2.8 sPzB 41)	90
7.5cm LG 40 Recoilless Gun (Airborne)	92
8.8cm Raketenwerfer 43 'Püppchen'	93
7.5cm Leichte Infanteriegeschütz 18	94
7.5cm Gebirgsgeschütz 36 (7.5cm GebG 36)	96
10.5cm leFH 18 (leichte Feldhaubitze)	98
10.5cm Gebirgshaubitze 40 (10.5cm GebH 40)	100
15cm sFH 18 (schwere Feldhaubitze 18)	102
15cm sIG 33 (Schweres Infanterie Geschütz 33)	104
21cm Mörser 18	106
15cm Nebelwerfer	108
28cm/32cm Wurfkörper-Spreng	109
30cm Raketenwerfer 56	111

Part IX - Anti-aircraft guns

Flak guns for mobile warfare	112
2cm Flak 38	113
2cm Flakvierling	114
Schwerer Wehrmachtschlepper (s.W.S.)	115
2cm Gebirgs-Flak 30	116
3.7cm Flak 36 and 37	117
Truck Platform for Light Flak	118
5cm Flak 41	119
5cm Flak 41 mit Sd.Ah 204	120
8.8cm Flak 18/36	121
12.8cm Flak 40 Zwilling	123

Part X – Vehicles

Supplying fuel to the armies of the Third Reich	125
Sd.Ah.S1	126
Raupenschlepper Ost (RSO)	127
NSU Kettenkrad 'HK 101' (Sd. Kfz.2)	128
Bison - 15 cm sIG 33 auf Fehrgestell Panzerkampfwagen II	129
Renault UE Chenillette (selbstfahrlafette Pak 36)	130
7.5cm Pak 40/4 auf Raupenschlepper (RSO)	131
Schwerer Wehrmacht-Schlepper (s.W.S)	132
15cm Panzerwerfer 42 (Sf.) auf LKW Opel 'Maultier' (Sd.Kfz.4/1)	133
Goliath (Leichte Ladungsträger Sd.Kfz.302)	134
Sd.Kfz.304 NSU 'Springer' Mittlere Ladungsträger	135
Borgward IV (Schwere Ladungsträger (Sd. Kfz.301 Ausf C)	136
'Wanse'	137
Kugelpanzer (Observation Tank)	138
Alkett VsKfz617 Minenräumer	139

Part XI – Late production munitions and miscellaneous kit

Infrarot-Scheinwerfer 1945	140
Fliegerfaust	141
X7 Rotkäppchen	142
Rheinbote Missile (Rhine Messenger)	143
7.3cm Propaganda-Werfer (Propaganda Rocket)	144

Part I – Small Arms
Mauser Karabiner KAR 98 – standard issue rifle

4x Zeiss ZF42 telescopic sight

Gewehrgranatengerät Rifle Grenade Launcher, Range 280m (306yds). See page 25 for full range of rifle grenades.

The Mauser Karabiner 98 Kurz was the standard issue rifle for all German forces in World War II. It was a five-shot bolt-action weapon, designed in 1935 from a long family line of guns going back to the 1870s. As a sniper weapon, it was fitted with telescopic sights; as a soldier's support weapon, it was fitted with the 'Schiessbecker' (shooting cup) grenade launcher, and in the last months of the war, an infrared night-scope. By Hitler's order, a programme to replace all 98Ks in front-line units with the MP44 assault rifle began when this was introduced in 1944.

Weight:	3.7kg (8lb 2oz)	Action:	Bolt-action
Length:	1,110mm (43.70in)	Range:	500m (550yds)
Barrel Length:	600mm (23.23in)		800+m (875+yds) telescopic sight
Cartridge:	7.92 x 57 Mauser	Loaded:	five rounds – stripper clip

MP 40 Maschinenpistole
Called by the British 'Schmeisser'

The first sub-machine gun (the MP 18) saw service in 1918, and during the intervening years underwent constant upgrades, from the MP38 through to the MP40. This weapon then became one of the iconic weapons of World War II. Designed by Heinrich Vollmer of the Erma Werks, (and not by Hugo Schmeisser, who had no involvement in its design) it went on to serve with all arms of service and on all fronts. With the arrival of the StG 44 in late 1943, a move began to standardise all the infantry weapons by rolling the rifle and machine pistol into one – the MP44.

Weight:	4kg (8.82lb)	Range:	70m (100m) (semi-automatic)
Length:	830mm (32.8in) extended stock 9mm	Rate of fire:	500rpm
Cartridge:	9mm Parabellum	Muzzle velocity:	400m/s (1,312ft/s)
Action:	Straight blowback		

Wimmersperg Spz-kr 7.92mm Assault Rifle
Concept design late 1945

The Wimmersperg Spz-kr Assault Rifle was designed to be manufactured using stamped metal sheeting. It used components mostly 'borrowed' from the German copy of the British Sten Gun, with the magazine and magazine release mechanism taken from the StG 44. This assault rifle never got beyond the planning stage and no weapon of this design saw service even in trials. However, its effect on the future design of modern weapons can be seen in the design of the British SA80 – 682 'BullPup', used by the British Army since 1980.

Designer:	Von Wimmersperg	Feed:	30 round magazine
Manufacturer:	Mauser, Fokker	Sights:	Iron, ZF-4 scope.
Effective range:	100-400m (440yds)		
Action:	Gas, select fire		

StG 44 (MP 44) (Sturmgewehr 44)
The world's first assault rifle

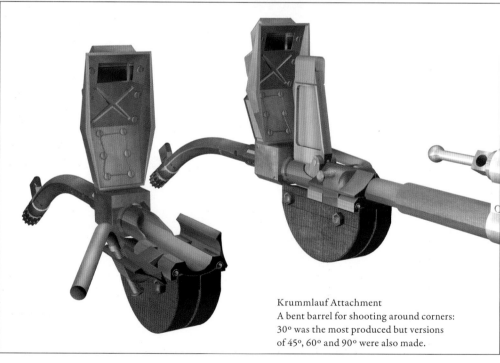

Krummlauf Attachment
A bent barrel for shooting around corners: 30° was the most produced but versions of 45°, 60° and 90° were also made.

The Sturmgewehr 44 (Assault Rifle 44) was the first mass-produced weapon designated as an assault rifle. Using mostly stamped metal parts, and lacking the fine finish of other German weapons, the Allies thought this weapon crude, poorly-made and cheap. It was only in post-war years that combining the conventional carbine, rifle and sub-machine gun in one general-purpose weapon, with resulting substantial cost savings, become apparent. This weapon was also the first to see combat in 1945 using the Zielgerat 1229 infrared night scopes (the Vampir).

Weight loaded:	5.13kg (11.3lb)	Action:	Gas-operated
Length:	940mm (37in)	Range:	300m; 600m (semi-automatic)
Barrel length:	419mm (16.5in)	Rate of fire:	550-600rpm
Cartridge:	7.92mm kurz	Muzzle velocity	685m/s (2,247ft/s)

PART I – SMALL ARMS

People's Weapons 1945
Late production small arms for the Volkssturm 1945

ERMA EMP 44.
Rejected for being too cheap and NOT looking like a 'real' gun. Bolt mechanism from the already 'tooled' MP40. Also 'borrowed' was the dual magazine loading tray, first designed for the MP40

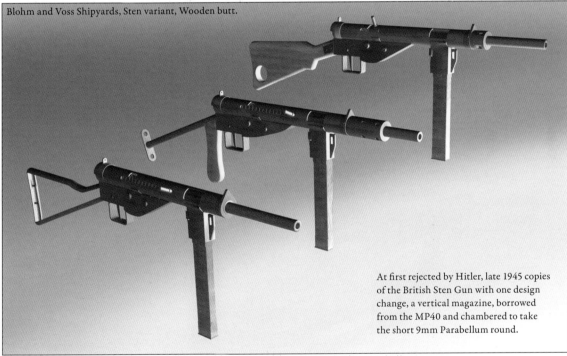

Blohm and Voss Shipyards, Sten variant, Wooden butt.

At first rejected by Hitler, late 1945 copies of the British Sten Gun with one design change, a vertical magazine, borrowed from the MP40 and chambered to take the short 9mm Parabellum round.

The need to mass-produce cheap guns with minimal 'tooling', which were simple to manufacture, led to borrowing designs from existing firearms. The need for a large supply of usable firearms to equip the Volkssturm (People's Army), supplementing those used by the armed forces, led to the creation of 'People's weapons'. Several versions of semi-automatic weapons and rifles were produced, often in little backstreet workshops, right up to the last day of the war. Many of these designs have influenced weapon designers right up to the present day.

People's Weapons 1945 – Volkssturmgewehr 1-5
Three weeks to design, three hours to manufacture

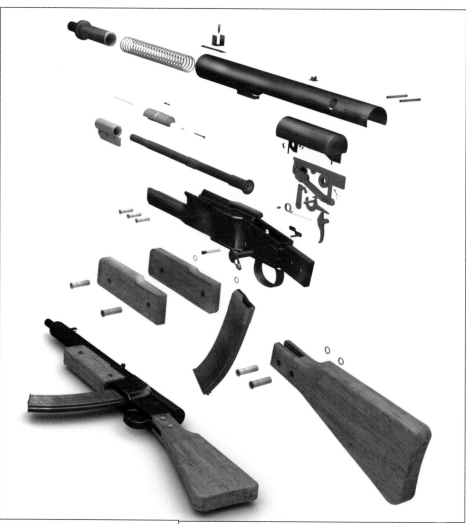

Built by the Gustloff-Werke/Suhl works and designed by chief designer Karl Barnitzke. After the war, he stated in an interview with American weapon investigators that the rifle took three weeks to design and three hours to manufacture. Over 10,000 were produced during the last months of the war, and were issued to Volkssturm battalions in 1945 as the Soviets marched into Germany.

Part II - Personal Equipment
Army Flare Gun
Flare gun, cartridge holder, cleaning iron and colour-coded cartridges

Heer flare gun, shown with different coloured cartridges.

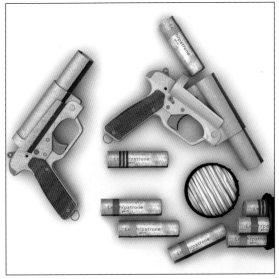

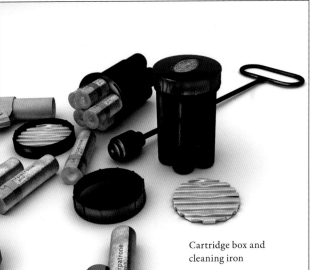

Cartridge box and cleaning iron

In these days of hi-tec battlefield telecommunications and night-vision headsets that come as standard, it is easy to forget that the flare was the only way to illuminate the night sky - with a parachute flare - and to speak to other units on the battlefield using designated 'colour of the day' flares. A range of flares were used, colour-coded for day or night, which in practice meant limited battlefield security. The flare gun was also used to fire small grenades.

A wide range of colours was issued, produced by many different manufacturers.

Model 1929 Army Flare Gun (with Sturmpistole Attachment)
Using the flare gun as a grenade launcher

Heer Model 1929, stripped down to show its component parts.

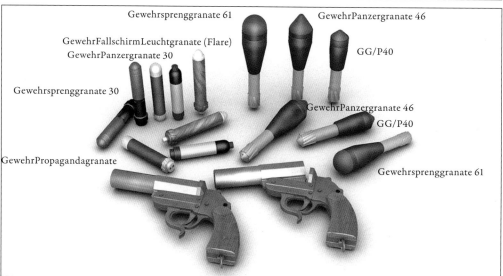

Gewehrsprenggranate 61
GewehrPanzergranate 46
GewehrFallschirmLeuchtgranate (Flare)
GewehrPanzergranate 30
GG/P40
Gewehrsprenggranate 30
GewehrPanzergranate 46
GG/P40
GewehrPropagandagranate
Gewehrsprenggranate 61

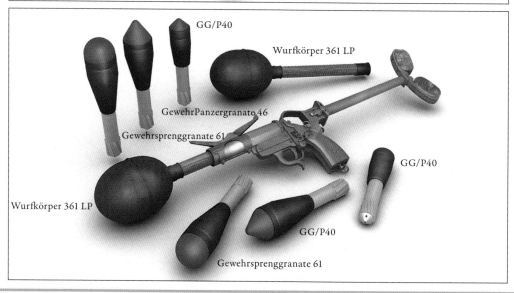

GG/P40
Wurfkörper 361 LP
GewehrPanzergranate 46
Gewehrsprenggranate 61
GG/P40
Wurfkörper 361 LP
GG/P40
Gewehrsprenggranate 61

A range of grenades was used, from high explosive to the shaped-charge for use against armour. The Sturmpistole attachment fired what was then the largest round, the Würfkorpor 361 LP. Other rounds could also be fired using this attachment.

PART II - PERSONAL EQUIPMENT

1945 Volkssturm Flare Gun
Crude but effective low cost flare gun, late war years

Issued to Volkssturm (*People's Army*) battalions.

These were 'end of war' flare guns, manufactured for the Volkssturm. Crude they might be, but they worked. They were made by using existing spares from different manufacturers and were assembled and built in small workshops.

A spring-loaded bolt was drawn back and compressed. When released, it fired a standard flare cartridge. Different colours signified different events: white for illumination, green, red and blue as colours of the day, or to indicate unit positions.

Telephone Wire Spool

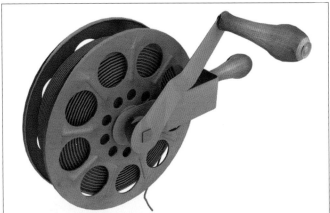

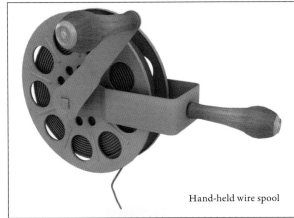

Hand-held wire spool

The German armed forces used three forms of telephone dispensing equipment, a hand-held version, the frame-mounted version (see top picture) and a vehicle-mounted reel. The frame-mounted spools gave 5km cable lengths, and were installed in a triangular steel framework. Leather straps were fitted, along with a cushioned pad, and the unit was worn like a rucksack. A leather pouch carried the winding handle, chain transmission, and other accessories. All the equipment was designed for ease of use in field conditions.

Alarmleuchtzeichen
Trip-wire flare

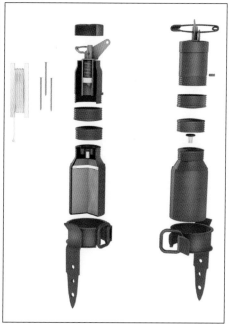

This was a standard issue 'trip-wire' flare. Also supplied with the flare unit was a lower 'stake' clip, which, with the nails supplied, could also double as a fitting to be used against a tree. The string, supplied in the same package, was used as a trip wire.

This was placed in front of defensive positions as an early warning device. In many variations, this simple device is still in use around the globe. The filling consisted of an illuminating compound.

Knijpkat, Dynamo Light
Non-Battery field lights/dynamo lights

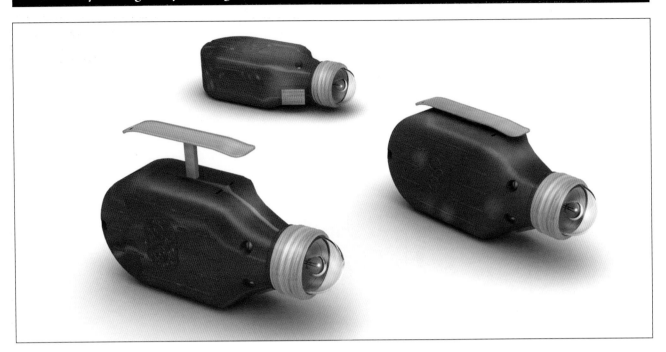

Knijpkat, made by Phillips of Holland, was a hand-pumped dynamo light. After the fall of Holland, Phillips began to produce this light for the German Armed Forces. Pumping the handle turned a flywheel that activated the dynamo and a small electrical current was discharged to light up the bulb.

This light, with many variations around the world, is still manufactured today.

PART II - PERSONAL EQUIPMENT

Field lights/torches

Early war years – small torch by the Daimon Company issued to the Wehrmacht

Early war years – large head torch by Daimon, Model No. 1511

Early war years – one of several designs used as an Officers' small torch, Model no. 8040

A wide and varied selection of torches was used by the Wehrmacht. Many officers privately purchased non-standard issue torches for their own use. All were battery-powered.

17

WORLD WAR II GERMAN FIELD WEAPONS AND EQUIPMENT

Dugout/Bunker - Wall or table light
Battery-charged light for use in temporary fortifications

Manufactured from recycled wax on a cardboard base, these candles, often called 'Hindenburg' candles, were very popular, giving both a little light and heat to warm liquids in the field. Simple paraffin lights for use in fortified field positions could also be seen. At top, a simple battery-operated wall or table light for use in bunkers and other fortified field positions, and a paraffin-fuelled lantern.

PART II - PERSONAL EQUIPMENT

Field Torch
Two versions of the same light by different manufacturers

By Daimon, Model No. 678

By Artes, Model No. 678

A standard-issued 'field torch' by two different manufacturers, Daimon and Artes. This battery-powered three-coloured (gel) light was issued to all ranks, and throughout all three arms of service. It was often worn hanging from a breast pocket. Photographs also exist showing Soviet troops using these captured lights in Berlin in 1945.

Part III – Grenades
Defensive and Offensive

For most of the war Germany used only a few types of grenade, mostly because of having vast stockpiles of the classic stick grenade left over from World War I. Another deciding factor was the vast amount of war booty captured during the war in the West and later during the fighting in the East. Only in the late war years, when a shortage of materials occurred because of the Allied blockade and air attacks (and later Allied advances into Germany from both East and West), did Germany look at different designs and cheaper means of manufacture. Eventually they improvised field weapons using whatever suitable materials were available. Concrete was one such material.

PART III – GRENADES

Model 34 Stielhandgranate
Defensive and offensive 'stick' grenades

Shown here, the classic stick grenade with the standard head. There is also an example with smooth and grooved slip-over fragmentation sleeves. A friction igniter was pulled to activate the 4½- or 7½-second delay fuse which would detonate the 0.89kg (1 lb 15½ oz) TNT charge.

Model 39 Eihandgranate 'egg' grenade
Also shown here with fragmentation sleeves

Early war years
Metal carrying case

Late war years
Wooden carrying case

A friction igniter was pulled to activate the fuse. Four types of fuse were used and these were colour-coded to indicate the length of the delay before detonation: yellow - 4½ seconds, red - 5 seconds, blue - 7½ seconds and silver grey - 10 seconds. Extra fragmentation sleeves could also be fitted to increase a grenade's destructive effectiveness.

Diameter of charge: 60mm
Height of container: 100mm

Model 42 Nelbel Eihandgranate
Smoke grenade 1942

Standard-issue smoke grenade. As the ring was pulled, it drew a wire through a friction igniter flash powder. The resulting flame set off the chemical filler, which burned its way through the paper plugs in the upper half of the casing, releasing the smoke. The split ring at the grenade base was to enable the item to be clipped to a soldier's webbing.

Diameter:	60mm	Filling:	170gm HC-Zinc 47/53gm
Height:	96mm/135mm	Material:	Pressed steel
Weight:	280 grams	Fuse/igniter:	Zunderschnuranzünder 29.39

WORLD WAR II GERMAN FIELD WEAPONS AND EQUIPMENT

Rifle Grenades

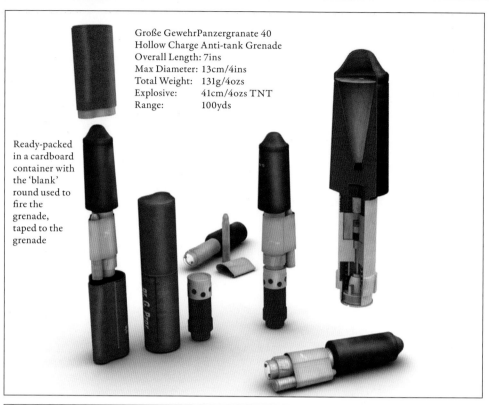

Große GewehrPanzergranate 40
Hollow Charge Anti-tank Grenade
Overall Length: 7ins
Max Diameter: 13cm/4ins
Total Weight: 131g/4ozs
Explosive: 41cm/4ozs TNT
Range: 100yds

Ready-packed in a cardboard container with the 'blank' round used to fire the grenade, taped to the grenade

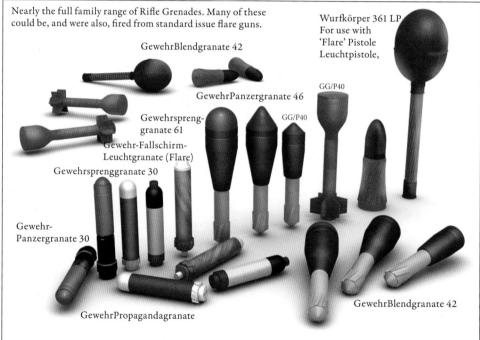

Nearly the full family range of Rifle Grenades. Many of these could be, and were also, fired from standard issue flare guns.

- GewehrBlendgranate 42
- GewehrPanzergranate 46
- Gewehrsprenggranate 61
- Gewehr-Fallschirm-Leuchtgranate (Flare)
- Gewehrsprenggranate 30
- Gewehr-Panzergranate 30
- GewehrPropagandagranate
- GG/P40
- GG/P40
- Wurfkörper 361 LP For use with 'Flare' Pistole Leuchtpistole,
- GewehrBlendgranate 42

A full range of offensive and defensive rifle grenades were produced by Germany, largely a hangover from World War I. In combat, however, it was soon clear that these small explosive charges were incapable of immobilizing – let alone destroying – the more heavily-armoured tanks or even armoured cars. Work soon began on a more deadly weapon, which would both standardise the supply chain and be capable of upgrading as demand required. In late 1942 the weapon arrived, in the shaped-charge warhead of the Panzerfaust 30.

Blendkörper (blinding device) (smoke) BK-2H/24
Glass smoke grenade

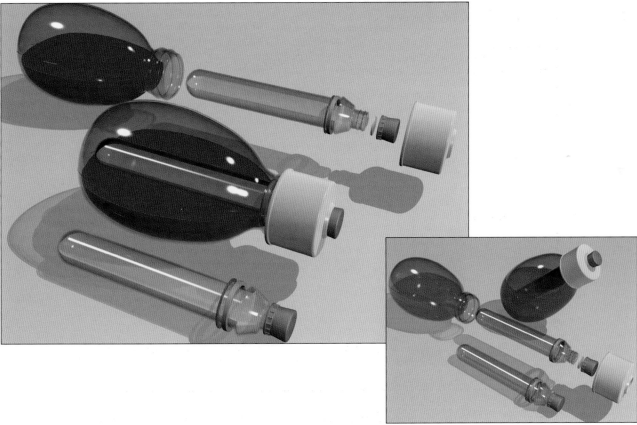

The glass smoke grenade was introduced in 1943. It consisted of a heavy glass bottle with a smaller inner glass vial. When thrown, the glass broke upon impact with a hard surface, causing the chemicals to mix and make smoke. It was used to blind an enemy or make a smoke screen. Over one million of this device were made between 1943 and the end of the hostilities.

Diameter:	2½in	Outer flask:	270 gm; titanium tetrachloride
Overall length:	4.8in	Inner vial:	36 gm; aqueous solution of calcium chloride
Weight:	17 oz		

Handgranate 343d

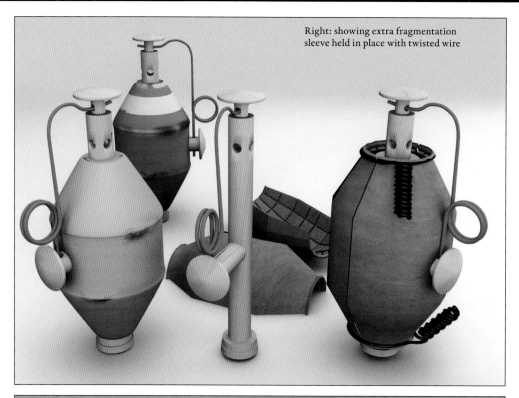

Right: showing extra fragmentation sleeve held in place with twisted wire

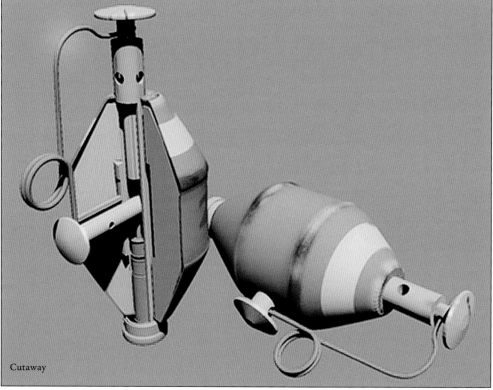

Cutaway

After the occupation of Denmark this grenade was adopted and given the new German nomenclature 'Handgranate 343d' (Offensive). Little is known of its use in the German forces. Some photographs have surfaced showing German troops putting fuses into this grenade but none indicate where the pictures were taken.

PART III – GRENADES

Late Production Munitions 1945 - Glass grenade

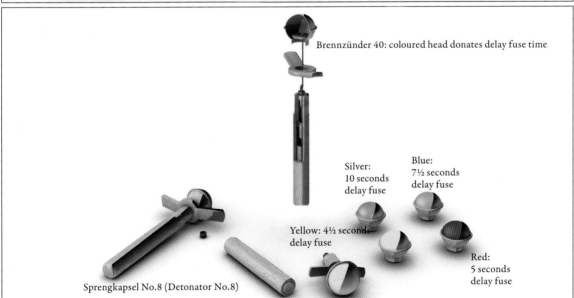

Brennzünder 40: coloured head donates delay fuse time

Silver: 10 seconds delay fuse

Blue: 7½ seconds delay fuse

Yellow: 4½ seconds delay fuse

Red: 5 seconds delay fuse

Sprengkapsel No.8 (Detonator No.8)

'Any material will do'. If it could be filled with an explosive, have a detonator attached to it and be thrown, it would 'do'. Concrete bomblets to Bakelite grenades, wooden mines to cardboard and porcelain mines – any material was used as the war went on. As a result of Allied bombing, then as the Allies advanced deeper into Germany, production became more difficult as the supply of weapons-grade materials began to dry up and factories were lost.

Improvised Concrete Grenade 1945
Late war years

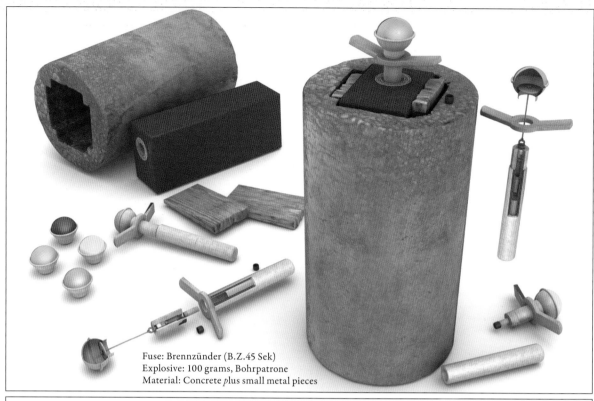

Fuse: Brennzünder (B.Z.45 Sek)
Explosive: 100 grams, Bohrpatrone
Material: Concrete plus small metal pieces

Fuse: Zündschnuranzünder 29
plus Zündschnur adaptor
Explosive: 100 grams, Bohrpatrone
Material: Concrete plus small metal pieces

As with field-improvised landmines, during the late war years, due to a lack of materials, many field improvisations of hand-thrown grenades were also made. The concrete grenade was one. The concrete was often mixed with metal scraps to increase injuries to enemy troops. A variety of charges and fuses were used.

Part IV - Hand-Held Anti-Tank Weapons
Panzerbüchse 39 (PzB 39) 1939
Anti-Tank Rifle

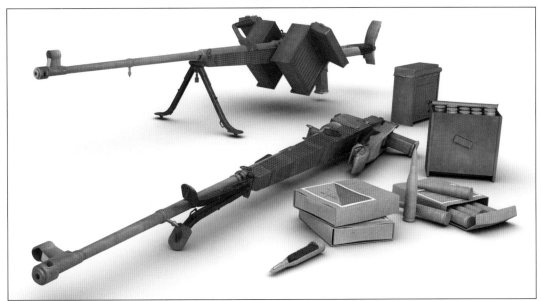

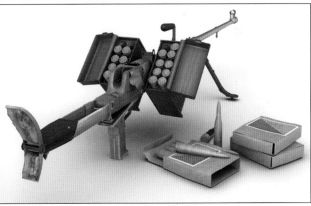

At the beginning of the war, all combatants had a large calibre high velocity rifle in the anti-tank role, however most armies soon relegated this weapon to history as tank armour increased beyond 25mm. Only the Soviet Union continued to use anti-tank rifles in the front lines. First conceived during World War I, to combat British tanks, development continued until in 1938 the Panzerbüchse 39 began to be issued to front-line units, seeing combat in Poland, 1939, when it came up against a Polish rival firing an anti-tank round with a tungsten core, which gave much better penetration, and was soon adopted by the Germans. As the war on the Eastern Front began to produce tanks with armour in excess of 100mm, this rifle was retired, only to return in an abridged form, sporting a shorter threaded barrel to allow the discharge cup for rifle grenades. Designated the GrB39 Grenade Launcher, it could fire a wide range of anti-tank and anti-personnel grenades to a range of 136 yards.

Calibre:	7.92mm	Range:	320m (350yds)
Muzzle Velocity:	1,265m/sec (4,150 ft/sec)	Weight:	12.60kg (27.78lbs)
Rate of Fire:	10 rounds per/m	Armour Penetration:	30mm at 100m (60°)
Feed:	Single shot	Barrel Length:	1086mm (42.75in)

Granatbüchse Model 39 (GrB 39) 1942
Grenade Launcher

Standard Cartridge: Wooden bullet
Effective Range: 136 yards+/-
Ammunition: Nearly all German rifle grenades

PART IV - HAND-HELD ANTI-TANK WEAPONS

Panzerfaust 30 (Klein)
Shaped-charge, hand-held anti-tank rocket

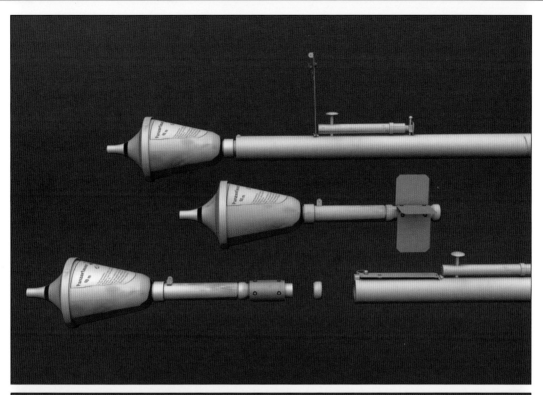

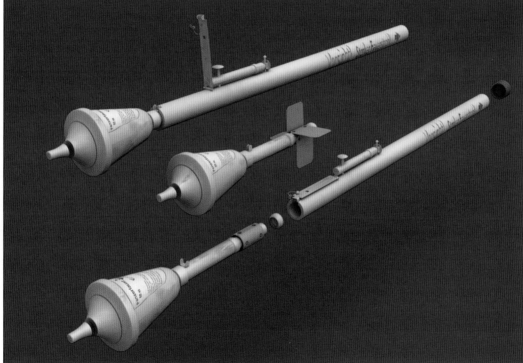

The world's first 'fire and forget' anti-tank weapon for the individual soldier was an idea attributed to Dr. Langweiler, an employee of HASAG (Hugo Schneider AG) (Leipzig). It was a small bomb, fired from a disposable tube launcher. A combination of recoilless gun and rocket motor propelled the bomb, stabilised by four spring steel fins, to its target. The range of this weapon was 30 metres. This shaped-charge could penetrate up to 140mm of armour at an angle of 30 degrees.

Weight:	9lb 4oz	Introduced:	Late 1943
Weight of charge:	3lb 4oz TNT		
Diameter of charge:	10cm		
Armour penetration:	30° 140mm		

Panzerfaust 60
Shaped-charge, hand-held anti-tank rocket

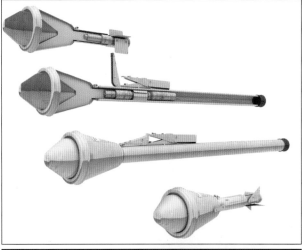
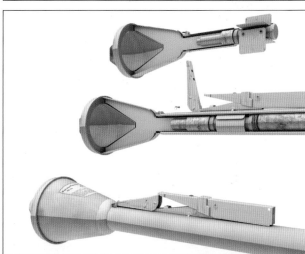
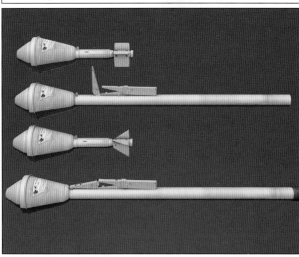
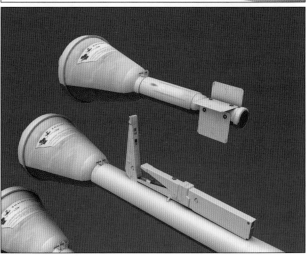

This weapon was fired from a disposable tube launcher, stabilised by four spring steel fins and propelled by rocket motor to its target. It was the deadliest of the Panzerfaust family of anti-tank weapons and the most widely used by all arms of the German forces. This weapon was ranged at 30, 60, and 100 metres. The 'shaped-charge' could penetrate up to 200mm of armour. It is seen here in its travel/supply crate.

Weight:	9lb 4oz	Introduced:	Late 1943
Weight of charge:	3lb 4oz TNT		
Diameter of charge:	10cm		
Armour penetration:	30° 200mm		

PART IV - HAND-HELD ANTI-TANK WEAPONS

Panzerfaust Family (30, 60, and 150)

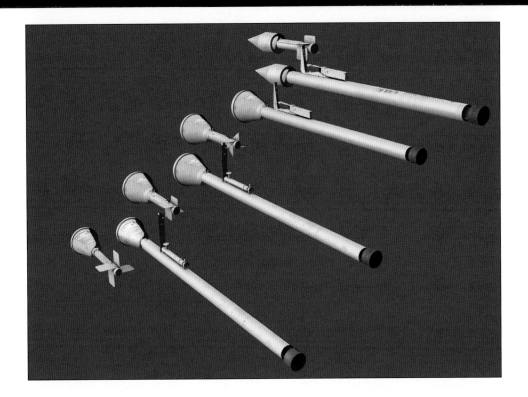

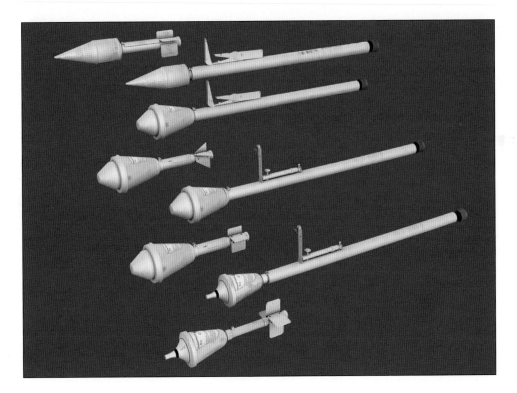

The full Panzerfaust family. The range of these weapons was 30, 60, and 100 metres. These 'shaped charge' 'bombs' could penetrate up to 200mm of armour and knock out most Allied armour of the period. Whereas in earlier campaigns, the tank took the lead, troops now had to lead to remove this threat to their tanks. RPGs and other modern 'fire and forget' weapons all have their origins here.

Weight:	9lb 4oz	Introduced:	Late 1942 onwards
Weight of charge:	3lb 4oz TNT		
Diameter of charge:	10cm		
Armour penetration:	30° 140-200mm		

Panzerfaust 150

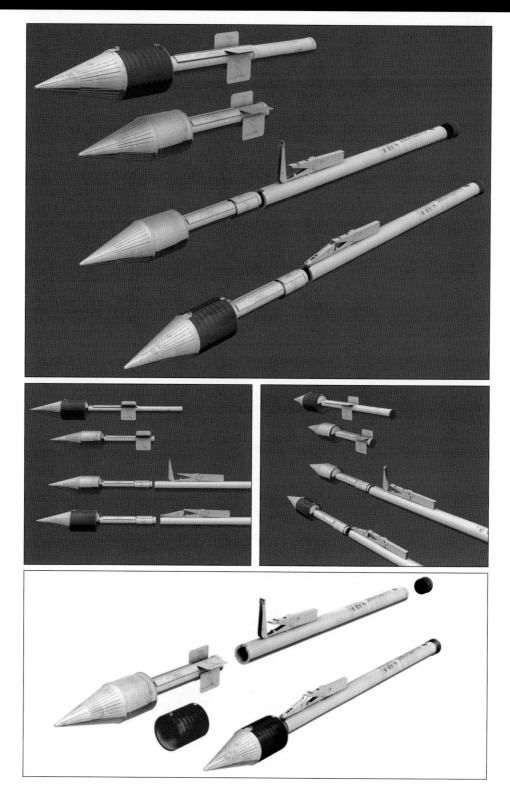

The production of the Panzerfaust 150 started late on in the war, so at its end few of them had entered service. The concept of the original Panzerfaust design remained the same but the warhead was smaller and more aerodynamically designed. It also had an increased range. This rocket bomb was designed to take a fragmentation sleeve for use as an anti-personnel bomb if required.

Weight: 9lb 4oz
Weight of charge: 3lb 4oz TNT
Diameter of charge: 10.15cm
Armour penetration: 30° 105mm (5.9 inches)

Introduced: 1945 onwards

PART IV - HAND-HELD ANTI-TANK WEAPONS

Panzerwurfmine (L)
Shaped-charge, hand-thrown, anti-tank bomb, used by 'tank killer' squads.

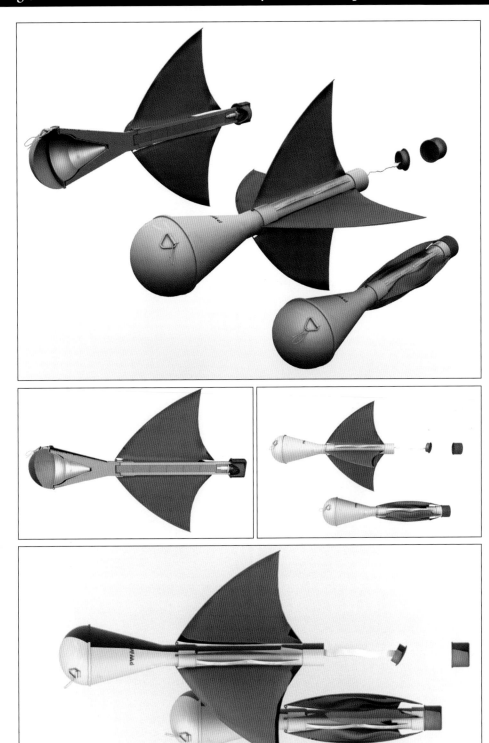

Thrown by swinging the bomb with an underarm movement, the canvas fins were released when the grenade was primed and guided the bomb though the air to its target. The Panzerwurfmine was an effective anti-tank weapon, with a maximum throwing range of 30 yards.

Weight:	1.35kg (9lb 4oz)	Length of fins:	279.4mm (11in)
Weight of charge:	(18½oz) RDX/TNT	Length of body:	228.6mm (9in)
Diameter:	114.3mm (4½in)		
Overall length:	533mm (21in)		

8.8cm Raketenpanzerbüchse 54/1 (Panzerschreck)
Shaped-charge, shoulder-fired anti-tank rocket

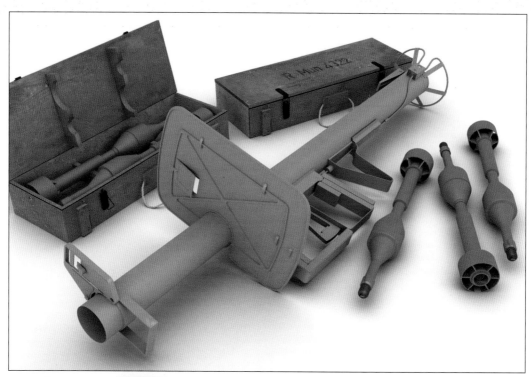

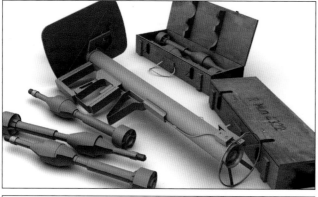

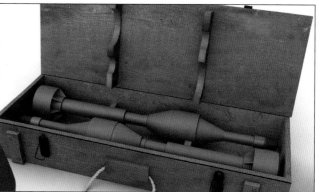

The first version of this rocket was a copy of an American bazooka captured in Tunisia in 1943. Sent to Germany for evaluation, it was hurriedly copied and issued to frontline units. The rocket-propelled grenade was enlarged form 2.36in. (US) to 8.8cm. The range of the first RP43 was about 165 yards. In 1944 a revised version was introduced, with an increase in range to 220yds. The fact that this weapon could now be fired without having to don protective clothing was a bonus for the user.

Weight:	1.35kg (9lb 4oz)		Length of fins:	279.4mm (11in)
Weight of charge:	(18½oz) RDX/TNT		Length of body:	228.6mm (9in)
Diameter:	114.3mm (4½in)		Range:	150m (165yds)
Overall length:	533mm (21in)			

Part V – Machine-Guns
Maschinengewehr 34 (MG 34)
The world's first general-purpose machine gun, 1934

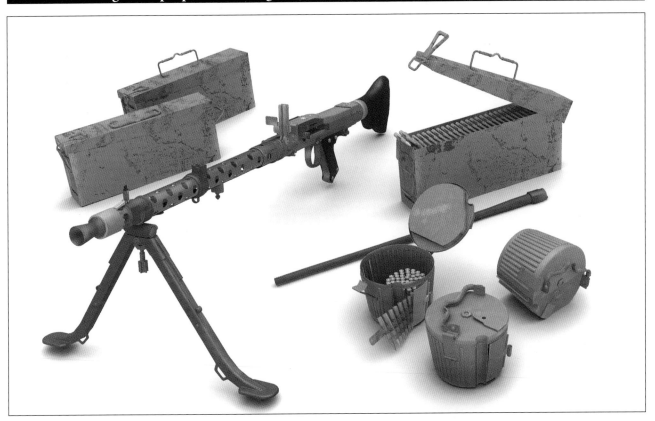

Introduced into service in 1934, this weapon was the most advanced machine gun of its day. It was light enough to be carried by one soldier and had a high rate of fire (800-900rpm). This weapon could also be fired on the move. It was belt-fed from a standard ammunition box (50-round belts x five, or from the small 50-round drum magazine). It was also used as secondary armament in tanks and other armoured vehicles, using a much thicker barrel to enable extended use before the barrel needed to be changed.

Designer:	Heinrich Vollmer
Manufacturer:	Mauser
Designed:	1934
Weight:	12.1kg (26.7lb)
With tripod:	19.2kg (42.3lb)
Rate of fire:	800-900rpm
Muzzle velocity:	762m/s (250ft/s)
Range:	Sights set at 2,000m
Tripod mounted	Up to 3,500m

Maschinengewehr 34 (MG 34)
The world's first general-purpose machine gun, 1934

In its role as a light machine gun, with a crew of two, the gun used a bipod and weighed only 12.1k (26.7lb). The ammunition was carried in boxes by the second crew member.

For fixed positions, a large tripod (Lafette MG34) was used. Extra-long 100-round belts were also available for extended firing, along with telescopic sights and a periscope so the gunner could see to fire while remaining under cover. The weapon also had special sighting equipment for indirect fire. The accurate range of the gun was also extended to 3,500m when mounted on the Lafette tripod.

This weapon is still in use in some parts of the world.

Designer:	Heinrich Vollmer	Rate of fire:	800-900rpm
Manufacturer:	Mauser	Muzzle velocity:	762m/s (250ft/s)
Designed:	1934	Range:	Sights set at 2,000m
Weight:	12.1kg (26.7lb)	Tripod mounted	Up to 3,500m
With tripod:	19.2kg (42.3lb)		

PART V – MACHINE-GUNS

Maschinengewehr 42 (MG 42)
7.92mm general-purpose machine gun - it influenced all post-war designed guns

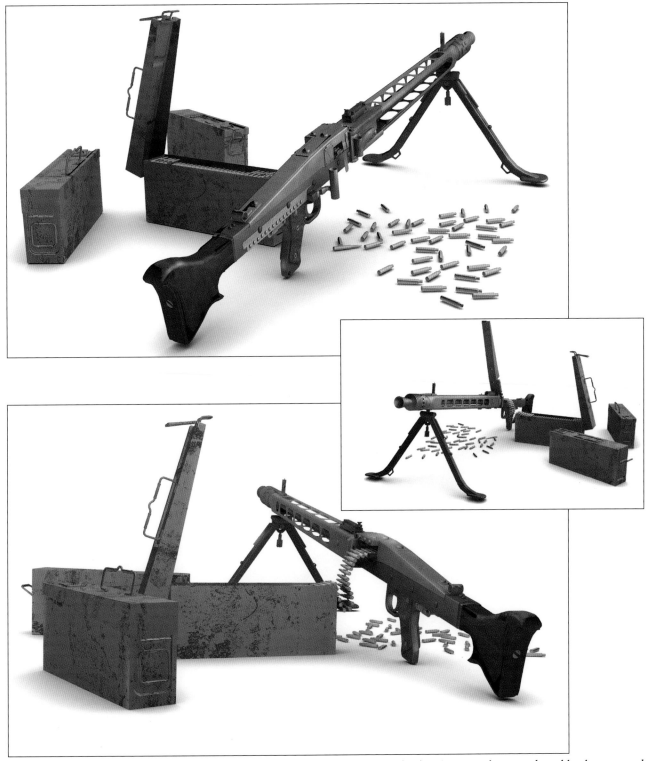

This gun entered service in 1942 and was meant to replace the MG34 completely. This never happened, and both were used right up to the end of the war. Its distinctive sound as it fired 1,200 to 1,500 rounds per minute was like ripping cloth.

This earned it a number of nicknames: 'Spandau' by the British (after the district where some were manufactured), 'Hitler's Buzzsaw' from the Americans and 'Linoleum Ripper' by the Soviets. Over 400,000 units were manufactured from late 1941 to the end of the conflict.

Designed:	1942	Muzzle velocity:	860m/s (2,822ft/s)
Manufacturer:	Mauser Werke AG	Cartridge:	7.92mm Mauer
Number built:	400,000+	Rate of fire:	900-1,500rpm
Weight:	11.57kg (25.51lb)	Effective range:	1,000m (1093.6 yds)
Length:	1,120mm (44 in)		

39

Maschinengewehr 42 (MG 42)
7.92mm general-purpose machine gun - it influenced all post-war designed guns

For fixed positions, a new tripod (Lafette MG42 - weighing 20.5kg) was designed. Extra-long 100-round belts (two connected belts of fifty rounds) were also available for extended firing, along with telescopic sights, and a periscope so the gunner could fire while remaining under cover. There was also special sighting equipment for indirect fire.

In a fixed position a crew of six were the norm, though in combat it was often reduced to three, gunner, loader and spotter.

Designed:	1942
Manufacturer:	Mauser Werke AG
Number built:	400,000+
Weight:	11.57kg (25.51lb)
Length:	1,120mm (44 in)
Muzzle velocity:	860m/s (2,822ft/s)
Cartridge:	7.92mm Mauer
Rate of fire:	900-1,500rpm
Effective range:	1,000m (1093.6 yds)

40

Part VI – Mortars
5cm 'Leichte' Granatwerfer 36
Standard mortar, early years

The mortar really came into its own during the static trench warfare of the Western Front throughout the years of World War I, as a means of placing a bomb into deep and narrow trenches. After that war all the main combatants designed new weapons. The German response was the 5cm Leichte Granatwerfer 36 (leGr36). Though issued to the army and used in combat it was considered to be a failure. Expensive to manufacture, and overly complicated to maintain and service, its short range and small charge led to its replacement in 1941 by the 8cm Granatwerfer 34 (8cm sGrW34).

Bomb:	0.9kg (2lb) (TNT)	Muzzle velocity:	75m/s (246ft/s)
Calibre:	50mm (1.97in)	Effective range:	50m (54.7yds) min
Elevation:	42° to 90°		510m (557.7yds) max
Traverse:	33° to 45°	Sights:	Telescopic, later deleted
Rate of fire:	15-30 rpm		

8cm Schwere Granatwerfer 34
Mortar, standard issue during the war

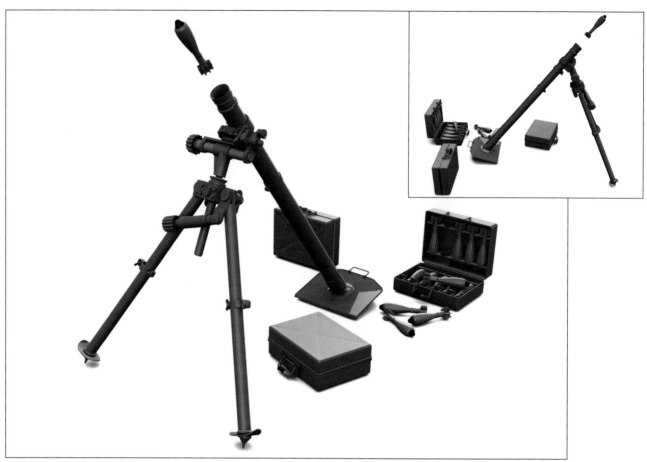

The 8cm Schwere Granatwerfer 34 replaced the 5cm Leichte Granatwerfer 36 in 1941 and soon became a weapon liked by all those who used it. The mortar was designed and manufactured by Rheinmetall-Borsig AG. Many German army units soon had two or three per platoon. Throrough training at all levels resulted in professional crews able to come into action quickly and retire from action equally speedily. The mortars were serviced by a crew of three, aimer, loader and armourer who fused the rounds, and fired H.E. air-burst and smoke bombs.

Bomb:	35kg (7lb 11oz) (TNT)	Rate of fire:	15-25 rpm
Calibre:	81.4mm (3.20in)	Muzzle velocity:	174m/s (571ft/s)
Elevation:	45° to 90°	Effective range:	2,400m (2,624yds)
Traverse:	10° to 23°		

Part VII - Mines And Demolition Charges
Improvised Anti-personnel Mines
Use of the Handgranate Stg24 and Druckzünder 35 (DZ35) fuse

Stick grenade cluster mine

Druckzünder 35 (DZ35), pressure igniter

Stick grenade anti-personnel mine made by using the 'head' of a grenade and a Druckzünder 35 (DZ35) pressure igniter

Field-improvised mines were made using the head from a stick grenade and pressure igniter Druckzünder 35. A 'Bangalore torpedo', used for blowing a path through barbed wire, was made from binding a number of explosive grenade heads together on a long stick or plank.

Explosive: 0.89g x number of heads
Fuse: Druckzünder 35 (DZ35)

'Plank' mine and Druckschiene for Teller 1 landmine
Improvised road blocks

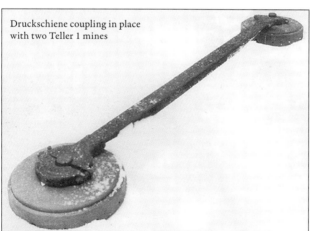

Druckschiene coupling in place with two Teller 1 mines

Road block using a Druckschiene coupling. The rope in the foreground is used to 'close the road' by dragging the coupled mines into place.

A Druckschiene or pressure bar as used with Teller mines. Anchored with ropes or chains to objects on either side of the road, this hasty construction used a pressure bar, attached by two half collars screwed together to fit tightly around the mantle of a Teller 1 mine. Up to six mines can be attached.

Plank mine:	3K demolition packs (as many as is required)
Materials:	Any wood planking available
Fuses:	Druckzünder 35 (DZ35) pressure igniter

'Plank' mine and Druckschiene for Teller 1 landmine
Improvised road blocks

'Plank' mine using 3K demolition packs

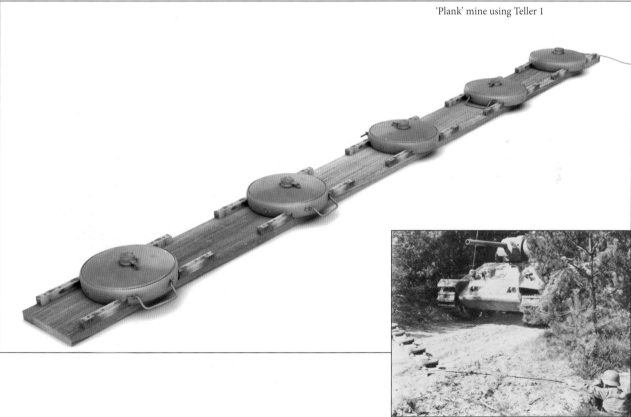

'Plank' mine using Teller 1

The plank mine served the same purpose as the manufactured Druckschiene, but was an improvised barrier using available materials. The top version used the 3K demolition pack with the Druckzünder 35 (DZ35) pressure igniter. The second version shown used Teller 1 mines wedged into position on a plank, guided into position by pulling a rope.

Holzmine 42
Wooden box mine

The mine consisted of a wooden box, made from ¾in. thick planks, divided by removable partitions into four compartments. The main charge was held in the two side compartments while the central area was for the 7oz primer charge. The front compartment contained a shearing flange secured to the wall by two ⅜in. wooden dowels. It had a central slot to take the end of the striker, and the igniter fuse sat on a small U-shaped block. A pressure of 200lb snapped the dowels securing the wooden shearing flange to the inner wall, which pushed the igniter (safety) pin out, freeing the spring-loaded striker to fire and set off the main charges. The fuse used was a standard Z.Z.42.

Diameter:	12 in.	Weight	18lb
Width:	¾ in.	Explosive:	Amatol 50/50
Deep:	12 in.	Materials:	Mostly wood
Height:	120mm	Fuse/igniter:	Z.Z.42

PART VII - MINES AND DEMOLITION CHARGES

S Mine
Defensive mine, three variants and travel/supply boxes

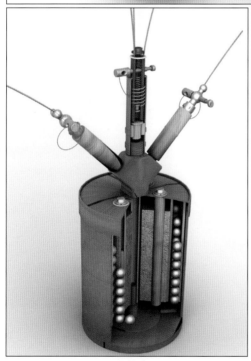

A German S-Mine, shown here with its early metal box and the wooden box used in later production. This mine contained 350 steel balls (diameter .372in) and two explosive charges. When the pressure-fuse was activated or a trip-wire snagged, the igniter discharged a flash of flame down the centre tube, firing the small first charge, which launched the mine into the air. At 6-7ft from the ground, the main charge exploded filling the surrounding area with steel balls up to a range of 150-200yds. The mine could also be fired in a controlled way by electrical methods.

Height (without igniter):	5 ins.	Material:	Sheet metal (zinc)
Weight of mine:	9lb 8oz approx	Fuse/igniter:	'Y' adaptor w/2 pull igniters
Weight of filling:	8oz to 1lb (TNT)		Pressure type
Diameter:	4in		Electric squib type

Brettstueck mine
Improvised plank mine

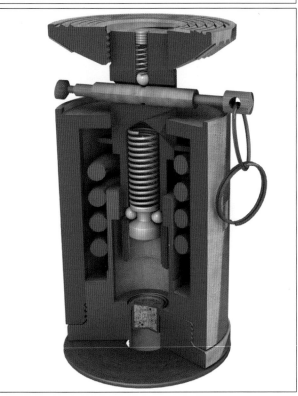

Behelfs-Brettstueckmine is the collective name for homemade mines of a similar design, used by units of the Wehrmacht. The chains were used to keep the plank on an even plane and stop any warping of the board as pressure was applied as a vehicle was driven over the mine. 1kg and 3kg Sprengbüchse demolition charges could be used.

Width:	1kg charge - 2.9ins	Material:	Sheet metal (zinc)
Height:	7.9ins	Fuse/igniter:	Druckzünder 35 (DZ-35)
Weight:	1kg.		
Explosive:	(PETN) TNT		

PART VII - MINES AND DEMOLITION CHARGES

Transport Rack for M-6 Teller Mine 42
Transportation rack used in early years of World War II for carrying two Teller mines: fuses were separate

This is a pre-war, all metal transportation rack for two Teller mines and a metal box for transporting a single mine. As the war progressed and materials became more difficult to obtain, manufacturers and suppliers began to adopt cheaper means for manufacture, as well as for labour and material. The result was often a good old-fashioned wooden crate or box

Teller Mine 42
Transportation box for carrying one Teller mine dating from middle years of World War II: fuses were separate

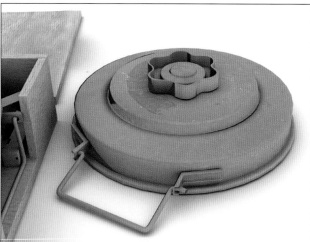

A Teller mine 42, seen here with its low-cost box. As the war went on, wooden boxes and wooden crates became the norm, replacing the more expensive pre-war metal boxes. This German mine was used in vast numbers throughout all the theatres of conflict. Often used as an anti-tank mine, with its Sprengkapelzünder fuse it was placed on an enemy tank by hand, often against the turret ring or turret overhang.

Weight of charge: 5.4kg
Weight: 9kg
Diameter of charge: 323mm
Height of charge container: 101mm

Fuse: A number of different fuses were used, subject to operational requirements

PART VII - MINES AND DEMOLITION CHARGES

Glass Mine
Glass mine with Hebelzünder and SF6 fuses, Sprengkörper 28 explosive charge

Topminenzünder 42 igniter (To.Mi.Z)

SF6 chemical igniter

A glass dish covered by a thin glass shear plate. When the shear plate was stepped on and broken, it crushed the top of the Buck (chemical) igniter or tripped the actuating lever of a Schücko igniter. The explosive charge was a standard German Sprengkörper 28 demolition block and a pressure of between 20-25lb was required to activate the mine.

Explosive:	Sprengkörper 28	Depth:	4.5in
Weight of charge:	Approx 7oz	Fuse/igniter:	Buck/Schücko
Pressure to detonate:	120-25lb		
Diameter of container:	6in		

Glass Mine
Glass mine with Hebelzünder and SF6 fuses, Sprengkörper 28 explosive charge

Topminenzünder 42 igniter

Hebelzünder igniter

A glass dish covered by a thin glass shear plate. When the shear plate was stepped on and broken, it crushed the top of the Buck (chemical) igniter or tripped the actuating lever of a Schücko igniter. The explosive charge was a standard German Sprengkörper 28 demolition block and a pressure of between 20-25lb was required to activate the mine.

Explosive:	Sprengkörper 28	Depth:	4.5in
Weight of charge:	Approx 7oz	Fuse/igniter:	Buck/Schücko
Pressure to detonate:	120-25lb		
Diameter of container:	6in		

PART VII - MINES AND DEMOLITION CHARGES

Panzerhohlladungsmine
Improvised anti-tank mine using a hollow charge warhead

Activated Hohlladungsmine with safety pin removed.

Hohlladungsmine with magnets removed.

This was a basic wooden box with a hinged lid (normally a nail was hammered in on both sides to act as the hinge). Inside was placed a Hohlladung 3kg anti-tank hollow-charge warhead, with its magnets removed. When pressure of approximately 100Ibs was applied to the lid, the unit was forced down, activating the Druckzünder-35 (DZ-35) igniter and firing the device.

Diameter:	170mm	Material:	Wooden box
Height:	225mm	Fuse/igniter:	Druckzünder 35 (DZ-35)
Weight:	5,000g		
Explosive:	1,500g		

53

Behelfs-Schützenmine S. 150
Anti-personnel mine

'Pot mine with the Druckzünder-35 (DZ-35) igniter

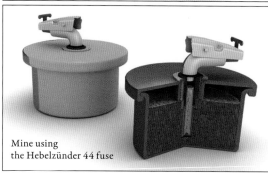

Mine using the Hebelzünder 44 fuse

Clearly shown in this cutaway (of the 'Buck' chemical crush igniter) is the ampoule filled with acid.

Behelfs-Schützenmine S. 150 anti-personnel mine, often called a pot mine. The body was made of pressed steel and held a 5¼oz explosive charge. A brass 'Buck' chemical crush igniter was the main fuse. When stepped on the body was crushed and a small glass ampoule filled with acid was broken. This caused a chemical reaction with a white powdered composition which surrounded the ampoule. This caused a flash which fired the mine.

Diameter:	2½in.	Material:	Pressed steel
Height:	2in.	Fuse/igniter:	A200 Buck
Weight:	12½ oz		
Explosive:	5¼oz powered picric acid		

PART VII - MINES AND DEMOLITION CHARGES

Panzerstabmine 43 (Pz.Stab.Mi.43)
Anti-tank mine

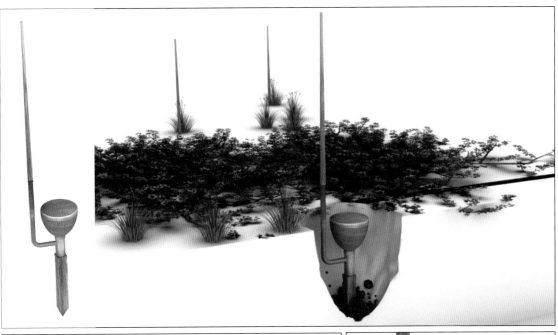

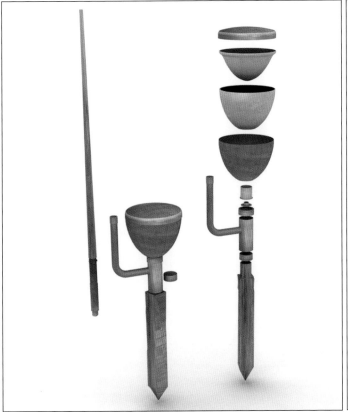

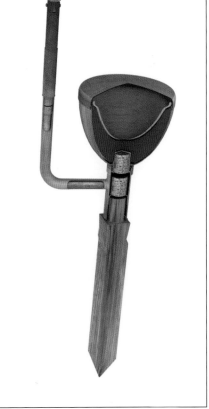

This was designed to destroy a tank and kill the crew, rather than disable an armoured vehicle. The mine was set in a rough hole about one metre large with tapered sides. When the tilt rod (91cm above the surface), was moved by a passing vehicle, the fuse was triggered and the explosion sent a super-heated jet against the weak floor of the tank. The heat and pressure killed the crew, and molten metal droplets often set fire to the ammunition. Before suspected corruption amongst manufacturers and Wehrmacht personnel stopped production, some 25,000 of these mines were made.

Diameter:	12.5cm	Material:	Metal
Height:	35mm	Fuse/igniter:	Knickzünder 43
Weight:	3kg		
Explosive:	800g (PETN plasticised)		

Shu-mine 1944
Anti-personnel mine

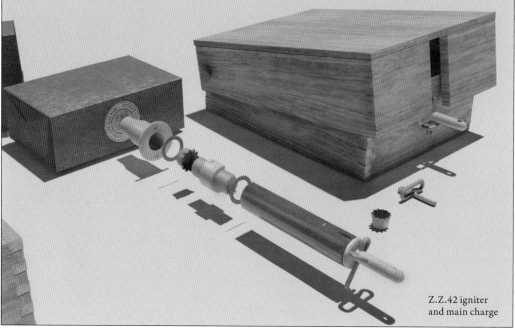

Z.Z.42 igniter and main charge

The mine consisted of a rough wooden box, made from plywood or another type of wood, which contained a demolition block fitted with a ZZ42 igniter and detonator. The box was covered by a hinged wooden lid held in place by two nails. Pressure on the lid pushed the safety pin out of the igniter, freeing the coiled spring and striker to detonate the igniter cap, detonator and explosive charge.

Weight of charge:	½lb TNT	Material:	wood/plywood
Width of charge:	3¼in.	Fuse/igniter:	Z.Z. 42
Length of box:	4¾in.		

PART VII - MINES AND DEMOLITION CHARGES

Engineers' tools for demolition, 1944

Cable-carrying drum

1kg blocks of Amatol explosives

Blasting machine

Crank for cable-carrying drum

Blasting Machine Glühzundapparat 37

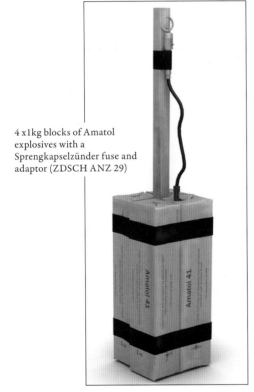

4 x 1kg blocks of Amatol explosives with a Sprengkapselzünder fuse and adaptor (ZDSCH ANZ 29)

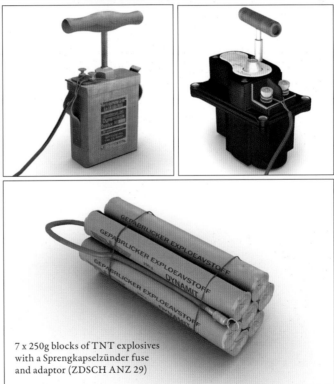

7 x 250g blocks of TNT explosives with a Sprengkapselzünder fuse and adaptor (ZDSCH ANZ 29)

By late 1944 Germany was practising the art of demolition, and pursuing a scorched earth retreat across the Eastern Front. Soldiers tried to burn or blast so as to leave nothing of use to the advancing Soviet forces. Livestock was rounded up and sent west. To slow the advance, they tried to destroy all bridges, road junctions and in the frozen wastes of the Russian winter, burn all habitats (a tactic used by all armies since the beginning of armed conflict). The sowing of minefields - even dummy minefields - and the use of booby-traps also aided the defenders wishing to slow the enemy's advance into Germany from the East and West.

Topf Mine
Late production anti-tank mine made up of cardboard, wood and glass

The underside of the mine clearly showing the glass screws and base-plate of glass

The glass igniter showing the two ampoules

A completely non-metallic anti-tank mine was made of pulped cardboard covered in pitch, with a glass chemical igniter to fire the main charge. A pressure of 300lb pushed the glass cap down, crushing the glass ampoules. One contained sodium and potassium and the other ethyl nitrate. The resulting flash fired the detonator, booster and main charge. It was impossible to find this mine with the mine detectors of the day.

Diameter:	12½ in.	Fuse/igniter:	Chemical
Height:	5½ in.		
Weight:	20lb.		
Material:	Pulped cardboard (pitch)		

PART VII - MINES AND DEMOLITION CHARGES

Topf Mine Porcelain 1945
Anti-personnel mine

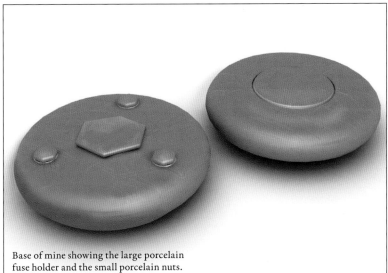

Base of mine showing the large porcelain fuse holder and the small porcelain nuts.

The glass igniter showing the two ampoules

Along with the manufacturing of glass, wood, cardboard and concrete landmines, it was only a matter of time before ceramics were used, and the first German porcelain mine were commercially produced. Both Japan and China had designed and manufactured small numbers of ceramic mines from the late 1930s onwards.

Wooden Topf Mine 1945
Late production anti-tank mine made up of cardboard, wood and glass

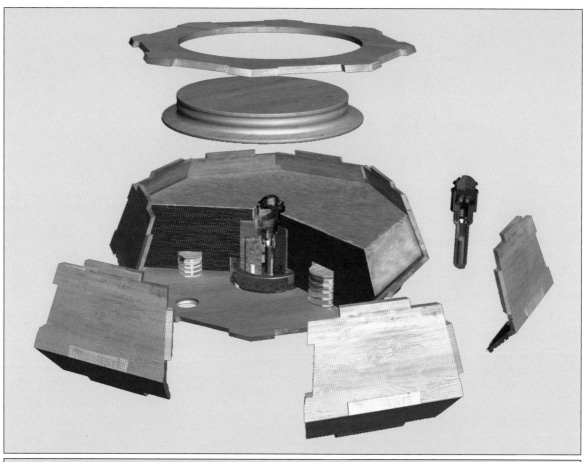

Base of mine showing the large glass fuse holder and the small glass nuts.

This was a completely non-metallic anti-tank mine. It was made of wood with a glass chemical igniter to fire the main charge. A pressure of 300lb pushed the pressure plate down, crushing the glass ampoules. One contained sodium and potassium and the other ethyl nitrate. The resulting flash fired the detonator, booster and main charge. It was impossible to find this mine with the mine detectors of the day.

Diameter:	12½in.	Fuse/igniter:	Chemical
Height:	5½in.		
Weight:	20lb		
Material:	Wood		

PART VII - MINES AND DEMOLITION CHARGES

Hohl-Sprung mine 4672 - hollow-charge bounding mine, 1945

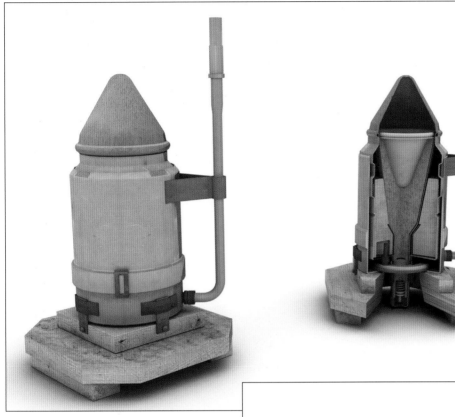

The mine consists of a Panzerfaust 60 shaped charge warhead held in a metal drum. The whole unit was mounted on a wooden board for stability, placed in the ground with only the Knickzünder 43 tilt rod showing. When the tilt rod was activated, the igniter fired the main charge, which threw the warhead against the underside of the tank where it exploded. Of the 59,000 manufactured, only three units of this mine are known to have survived the war. This is mainly because they were destroyed in situ as they were fitted with anti-lifting devices.

Diameter:	158mm		Material:	Metal
Height:	285mm		Fuse/igniter:	Knickzünder 43/Kippzünder
Weight:	10kg			
Explosive:	1,587g (PETN) RDX/TNT			

Panzermine A1 - Aluminium Type
Lightweight anti-tank mine

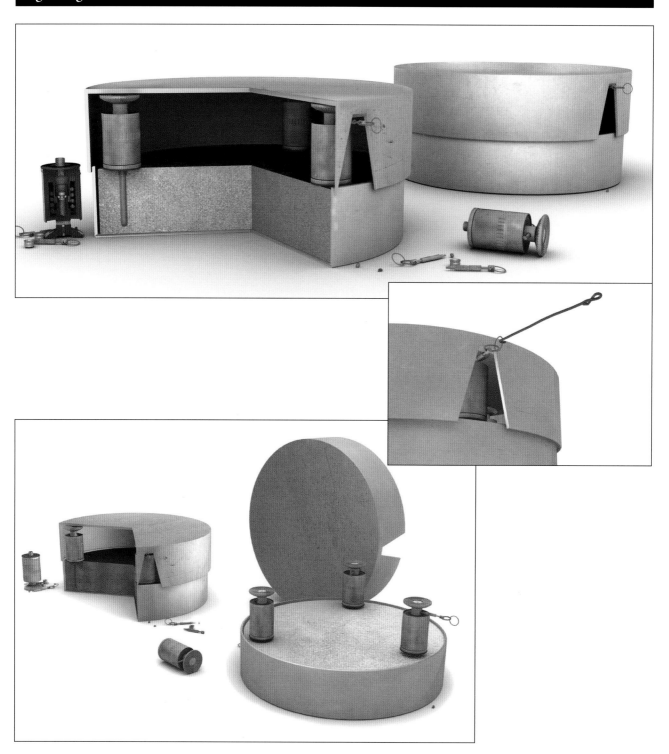

This was a lightweight anti-tank mine made of aluminium. The lower half held 3,500g of explosive, activated by 180-160 kg of pressure on any of the three DZ-35 igniters. Once placed in the ground the safety pin is withdrawn through the side opening in the top half. The lid was then carefully rotated to access the next safety pin and so on till all three igniters were armed. The mine was then camouflaged by a covering of earth, leaves etc.

Diameter:	300mm	Material:	Aluminium
Height:	120mm	Fuse/igniter:	Druckzünder 35 (DZ-35)
Weight:	6,400g		
Explosive:	3,500g		

PART VII - MINES AND DEMOLITION CHARGES

Panzermine Steingüt (Pz.Mi Steingüt) Clay Type
Late production mine made of baked clay

Very difficult to find information on this mine. This is very much an artist's impression of the device drawn from a number of sources.

This mine was made from baked clay (sometimes referred to as an 'earthenware mine'). The lower pot holds both the explosive and two detonators, Type ZZ-42. The pressure of around 70-80 pounds caused by a man stepping on the mine pushed the lid down, thereby ejecting the safety pin from the detonator(s). The detonator was attached to detonation cord, which then fired the igniter attached at the end. The mine's main charge was then detonated. As the mine had few metal parts, it was very hard to detect with the mine detectors of the period.

Diameter:	215mm	Material:	Baked clay
Height:	254mm	Fuse/igniter:	Two ZZ-42
Weight:	10,000g		
Explosive:	7,500g		

Concrete Block Mine
Improvised anti-personnel mine

Fuse: Z.Z.42, pressure igniter
Explosive: Sprengkörper 28
Material: Concrete plus small metal pieces

Fuse: Druckzünder 35 (DZ35), pressure igniter
Explosive: 0.9kg, 5cm mortar bomb
Material: Concrete plus small metal pieces

In the later war years, a lack of materials meant many impromptu devices being made. One field improvisation was to construct, using available materials, a concrete mine. Tied to a tree with a tripwire, placed on a wooden stake, or buried in the ground, the concrete was often mixed with metal scraps to increase injuries to attacking troops.

PART VII - MINES AND DEMOLITION CHARGES

Riegelmine 43 (R.Mi.43), Sprengriegel 43 (Spr.R.43)
Anti-tank high explosive bar mine

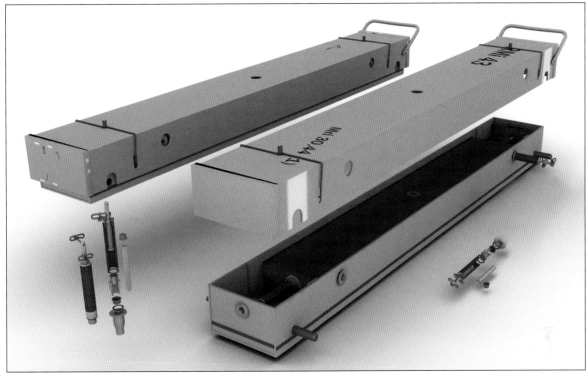

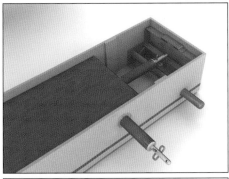

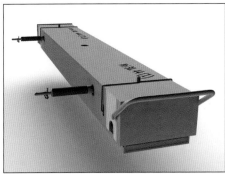

The Sprengriegel 43 high-explosive bar mine was laid in rows across open country or roads. It was constructed of sheet metal. A tray base, which had shear wires threaded through to the inner sides, contained the main explosive. This was a metal-encased charge of 8.8lb of TNT. A lid acted as the pressure plate. The charge was provided with five igniter sockets: two for the main igniters (ZZ 42), and three for the anti-lifting or trip-wire igniters. Some 3,051,400 were made and issued during the war. The mine was based on an Italian bar mine design.

Overall length:	31½in (800mm)		Material:	Sheet metal
Overall width:	3¾in (95mm)		Fuse/igniter:	Two x ZZ-42
Height (laid):	3½in (90mm)		Pressure (at ends):	440lb
Explosive:	8.8lb TNT		Pressure (centre):	880lb

Main fuses/Igniters

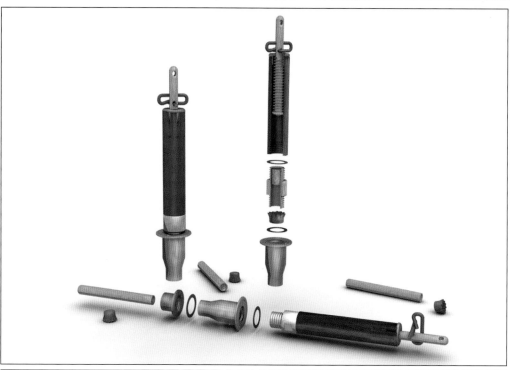

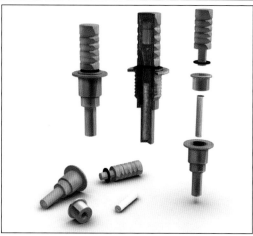

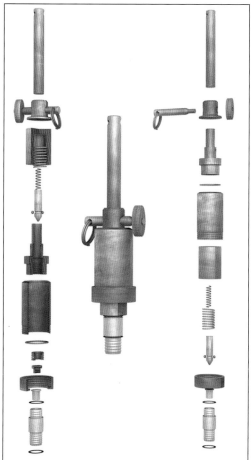

Brennzünder 40: coloured head indicates delay fuse time
Blue: 7½ seconds delay fuse
Red: 5 seconds delay fuse
Yellow: 4½ seconds delay fuse
Silver: 10 seconds delay fuse

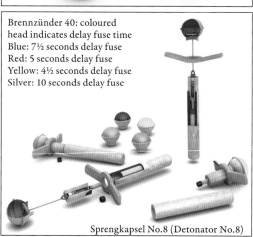

Sprengkapsel No.8 (Detonator No.8)

PART VII - MINES AND DEMOLITION CHARGES

Main fuses/Igniters

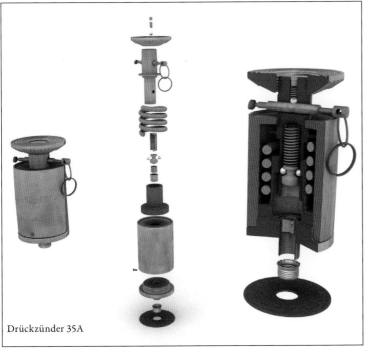

Drückzünder 35A

100 grams
Bohrpatrone 28

Panzerschnelmine A and B
Wooden anti-tank box mine

Similar to the 'Shu-mine', this larger version used 6 kilos of picric acid wrapped in waterproof paper as its main charge. The initiating charge weighed 200g. All the material was placed in a crude wooden box. Two simple nails acted as the hinge for the angled lid. Version A used a ZZ42 fuse, which, when the lid was pushed down forced the safety pin out, releasing the spring to fire the detonation charge. Version B used a brass 'Buck' chemical crush igniter as the main fuse. When driven over, the body was crushed and a small glass ampoule filled with acid was broken. This chemical interaction with the white powdered composition surrounding the ampoule caused a flash that fired the mine.

Weight of charge: 6kg picric acid in waterproof paper
Fuse: Buck (chemical fuse)

PART VII - MINES AND DEMOLITION CHARGES

Behelfmine E.5
Tin box mine using French munitions

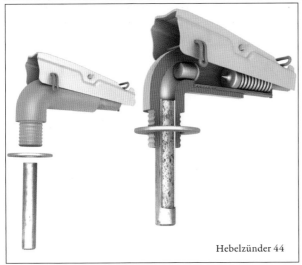

Hebelzünder 44

A simple improved anti-personnel mine could be made using a simple tin box, and, from the vast quantity of captured French war material, four French defensive grenades and one offensive grenade. These were placed in the centre of the box and fused using a Hebelzünder 44 attached to a trip-wire. Other fuses could also be used. Most German fuses would fit most mines. All that was needed was the right adaptor. The 'f' in the name simply signifies a French munition being used.

Weight of charge:	60g x 5	Fuse:	A number of different fuses were used, subject to operational requirements
Weight of grenade:	600g x 5		
Diameter of tin:	14.5cm		
Height of container:	8.25cm		

WORLD WAR II GERMAN FIELD WEAPONS AND EQUIPMENT

1kg and 3kg Demolition Charges

3kg and 1kg Demolition charges, seen here with just a few of the types of fuse used to detonate these devices.

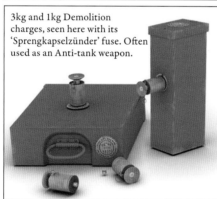

3kg and 1kg Demolition charges, seen here with its 'Sprengkapselzünder' fuse. Often used as an Anti-tank weapon.

Druckzünder 35A fuse

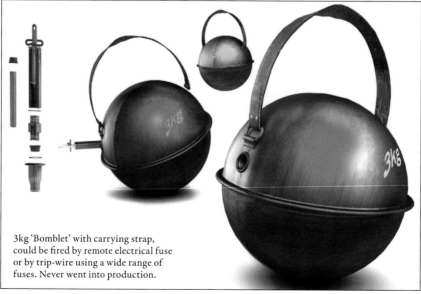

3kg 'Bomblet' with carrying strap, could be fired by remote electrical fuse or by trip-wire using a wide range of fuses. Never went into production.

Weight:	3.5kg
Weight of charge:	3kg (amatol)

PART VII - MINES AND DEMOLITION CHARGES

Flaschenmine 42
Glass bottle mine

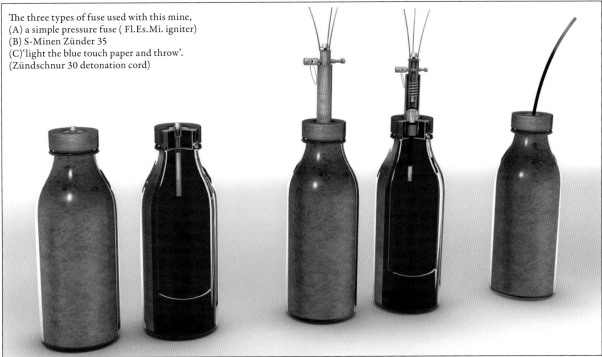

The three types of fuse used with this mine,
(A) a simple pressure fuse (Fl.Es.Mi. igniter)
(B) S-Minen Zünder 35
(C) 'light the blue touch paper and throw'.
(Zündschnur 30 detonation cord)

Placed in concrete and mixed with metal scraps or (as above) buried as a mine. Also fired by use of trip-wires

'Light the blue touch paper and throw'. Modified as a grenade using Zündschnur 30 (detonation cord) as the detonator charge

Fuse: A number of different fuses were used, subject to operational requirements.

Improvised Concrete Stake Mine
Late war field improvisation

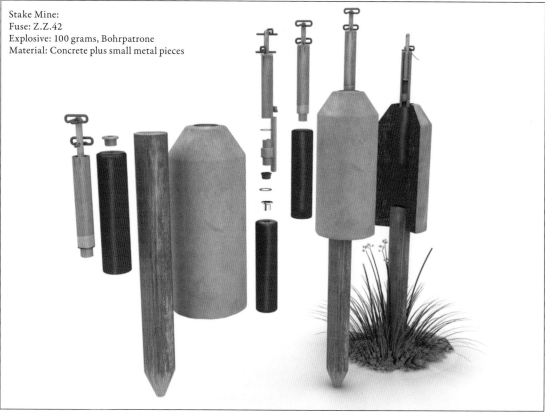

Stake Mine:
Fuse: Z.Z.42
Explosive: 100 grams, Bohrpatrone
Material: Concrete plus small metal pieces

Many field improvisations were made in the later war years because of a lack of materials. The concrete mine was one. Wired to a tree with a tripwire or placed on a wooden stake, the concrete was often mixed with metal scraps to increase injuries to attacking troops. The basic concept (IED - Improvised Explosive Device) is still used today in conflicts around the globe.

Kippzünder 43 Tilt Switch (K.i.43)
Pivoting tilt fuse

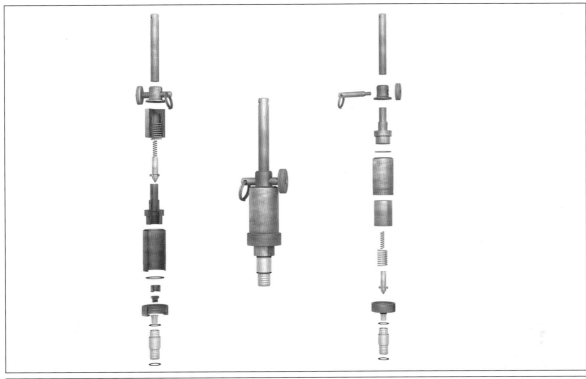

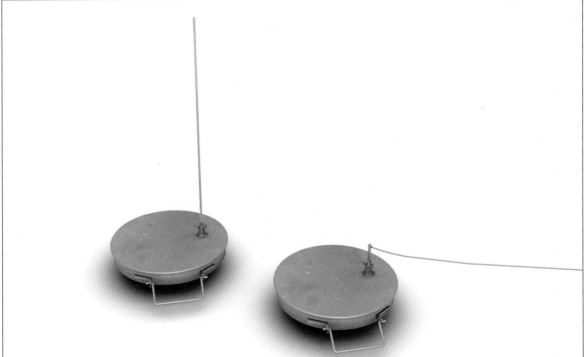

Introduced in 1943, this was a simple fuse, and only needed a pressure of between 15 and 25lb, or 1½lb if the tilt rod extension was used. Movement in any direction activated the device and fired the detonator, which in turn exploded the mine. It was set as a booby-trap for ground troops when attached to anti-personnel mines or laid in the paths of possible tank movements attached to anti-tank mines. From the point of view of a tank driver, looking for anti-tank guns, the rod was almost invisible.

Teller mine Booby-trap
Use of the Teller mine with anti-lifting devices

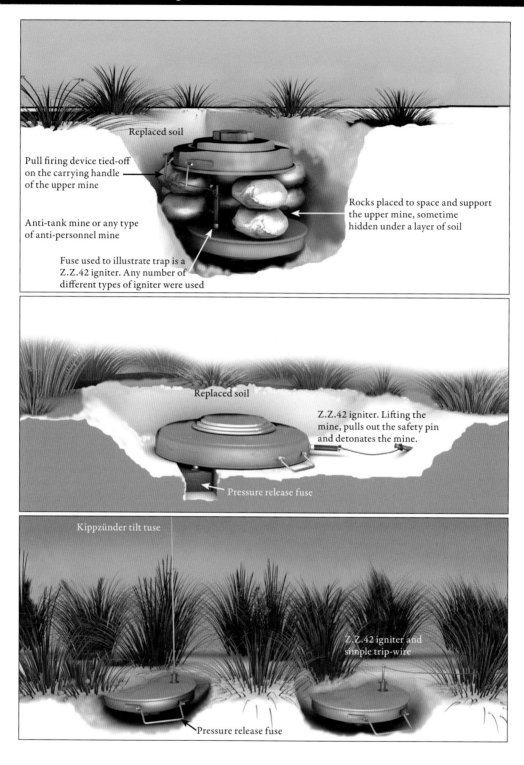

With the introduction of anti-lifting devices (mostly pressure release types), Allied troops often had to destroy mines fitted with them *in situ* as it became too dangerous to try and remove them when clearing a minefield. Booby-trapping some mines in a sowed field would lead to a slower clearance of the minefield, while also causing casualties amongst troops and mine-clearance engineer units.

PART VII - MINES AND DEMOLITION CHARGES

Entlastungszünder 44
Anti-lifting device

Introduced in 1944, the Entlastungszünder 44 was a simple yet deadly addition to the art of mine-laying. It was a cheap, clockwork anti-lifting device, easily produced in backstreet workshops. A spring cap, held under pressure by the weight of the mine placed on top of the device, was activated when the pressure was released by removing the top mine. The clockwork mechanism then fired the detonator and booster charge, which detonated the device's main charge. This explosion then fired the top mine, causing casualties amongst troops clearing the minefield. The priming charge was 100g of HE.

Haft-Hohlladung 3kg
Shaped-charge, hand-held anti-tank mine, used by 'tank killer' squads

Held in place against the side of a tank by three strong magnets, a friction igniter was pulled to activate the 4½- or 7½-second delay fuse, which would detonate the 0.89kg (1lb 15½oz) charge of TNT. It was issued to special-purpose engineer teams and 'tank killer' units. It could penetrate 100mm of armour.

Weight:	3.49kg (7lb 12oz)
Weight of charge:	0.89kg (1lb 15½oz) TNT
Diameter:	180mm (4½in.)
Height of charge container:	197mm (7½ in.)
Height of magnets:	70mm (2½in.)

PART VII - MINES AND DEMOLITION CHARGES

Stielhandgranate 24, 39, 43
Shaped-charge, hand-held anti-tank mine, used by 'tank killer' squads

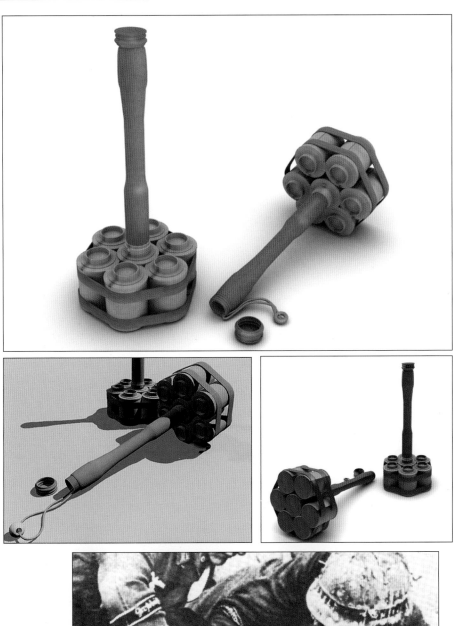

Building an Anti-tank mine by wiring 'spare heads' together

This consisted of a standard stick grenade with six explosive heads wired together, giving a combined charge of 42/49oz. This was capable of damaging most Allied tanks. The combination was also used by engineer demolition teams, and was known as Geballet Ladung.

Weight of charge: 0.89kg (x6) 42-49oz TNT

Panzerhandmine 'Sticky mine'
Shaped-charge, hand-held anti-tank mine, used by 'tank killer' squads

Held in place against the side of a tank by a strong sticky substance, a friction igniter was pulled to activate the 4½- or 7½-second delay fuse, which would detonate the charge. The mine was issued to special-purpose engineer teams and 'tank killer' units, though it was generally disliked and was soon replaced by the Haft-Hohlladung 3kg.

Panzerhandmine 3 Magnetic Mine
Shaped-charge, hand-held anti-tank mine, used by 'tank killer' squads

Held in place against the side of a tank by three strong magnets, a friction igniter was pulled to activate the 4½- or 7½-second delay fuse, which would detonate the charge. It was issued to special-purpose engineer teams and 'tank killer' units.

PART VII - MINES AND DEMOLITION CHARGES

Improvised Anti-tank Variants
Field improvisations using mines and demolition charges

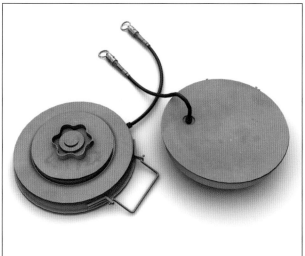

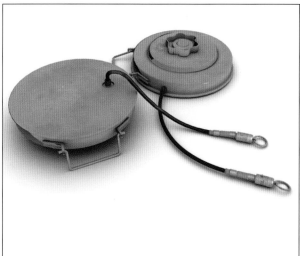

Seen here with their 'Sprengkapelzünder' fuse and adaptor (ZDSCH ANZ 29) are 3kg Sprengbüchse and 1kg Sprengbüchse demolition charges. When the ring is pulled, a wire is drawn through a capsule containing flash powder. The ignition of this highly combustible substance sets fire to the detonation cord that runs from the fuse to a No 8 detonator (Sprengkapel No 8), which fires the main charge. In its anti-tank role the device needed to be placed on the vehicle or under a track to be effective. The same applied for the Teller mine.

Part VIII - Artillery and crewed anti-tank weapons
3.7cm Pak 35/36 (Panzerabwehrkanone 36)
Until 1942 this was the main anti-tank gun of infantry units

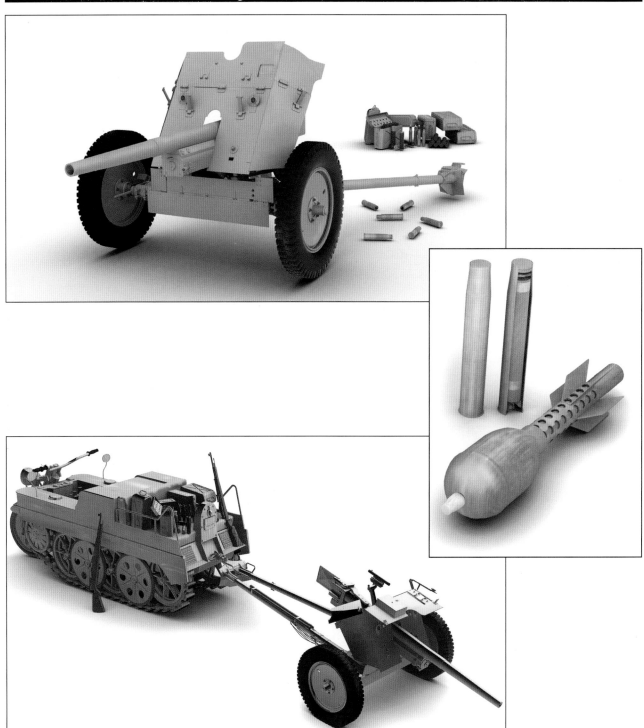

First issued to the German army in 1928 as a horse-drawn gun, this weapon was later upgraded so it could be towed by motorised transport. The upgrade meant changing the wooden wheels for magnesium alloy wheels with pneumatic tyres. The opening of the Ost Front in 1941, when the gun met the KV-1 and the T-34, meant it was rendered ineffective overnight. A massive programme was undertaken to replace it with the 7.5cm Pak 40, though it remained in service as an anti-tank gun until mid-1942. Many were then put on half-tracks to supplement firepower.

Designer:	Rheinmetall	Elevation:	-5° to +25°
Rate of fire:	13rpm	Traverse:	30°
Muzzle velocity:	762m/s (2,500ft/s)		
Range:	300m (328yds)		

PART VIII - ARTILLERY AND CREWED ANTI-TANK WEAPONS

3.7cm Pak 35/36 (Panzerabwehrkanone 36)
Until 1942 this was the main anti-tank gun of infantry units

The introduction in 1943 of the Stielgranate 41 shaped charge gave the Pak 36 a new lease of life. The hollow tube of the projectile was fitted over the barrel from the front, and was fired using a blank cartridge. Even with its low velocity and limited range, this shaped charge could defeat any tank of the period. Its ease of handling and mobility made it a very welcomed addition to the firepower of the units to which it was issued – these included Fallschirmjäger units.

Stielgranate 41:		Max range:	800m (870yds)
Weight:	8.6kg (19lb)	Warhead:	2.42kg (5.3lb) TNT
Muzzle velocity:	119mps (361fps)		
Projectile range:	(328yds)		

5cm Pak 38 L/60 (1940)
Anti-tank Gun

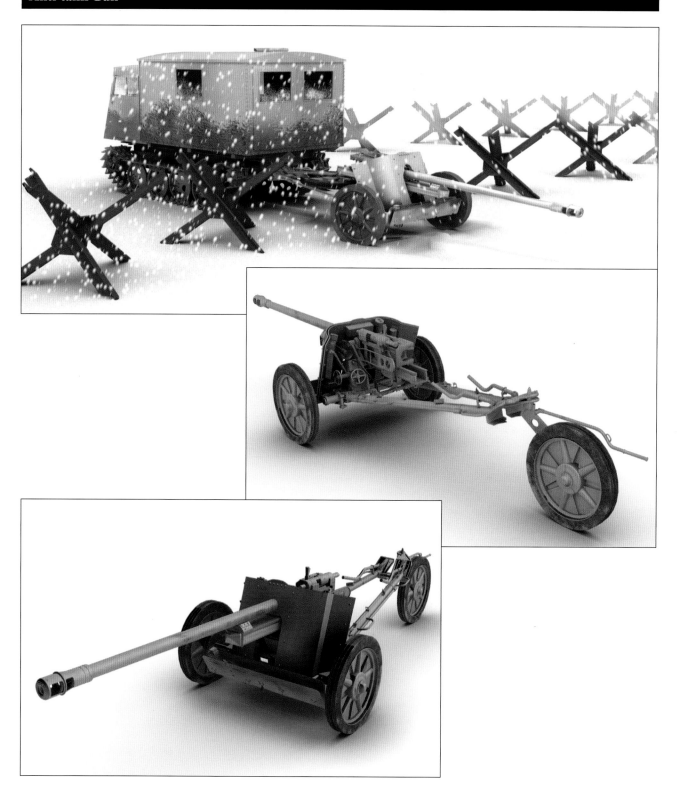

Designed in 1938 by Rheinmetall-Borsig, it entered into service in late 1940. At the time of the invasion of Russia, it was the only gun that could defeat the KV-1 and the T-34. Firing the tungsten-cored AP40 round, this ammunition could penetrate the armour of all Allied tanks during the war's middle years. Due to its construction. the gun was light for it size and easy to handle, its third wheel giving it an easy ride over rough ground. Later, as a result of battlefield conversions, this weapon soon

Calibre:	50mm (1.97in)	Muzzle Velocity:	1198 m/s (3930 ft/sec)
Traverse:	65°	Armour Penetration:	A.P. 61mm at 500 yards (30°)
Elevation:	-8° to +27°	Armour Penetration:	A.P.40 86mm at 500 yards (30°)
Shell Weight:	A.P. 2.25kg (4lb 15¼ oz)	Complete Weight:	986kg (2,174lb)
Shell Weight:	A.P.40 0.97kg (2lb 2¼ oz)		

PART VIII - ARTILLERY AND CREWED ANTI-TANK WEAPONS

5cm Pak 38 L/60 (1940)
Anti-tank Gun

found itself mounted on a wide range of self-propelled vehicles, both wheeled and tracked. It was also given an automatic feed (BK5) and mounted in a number of aircraft types for use as an airborne anti-tank 'buster'. Due to increases in armour and battlefield assessment, the plan was to replace the 5cm Pak with its more powerful successor, the 7.5cm Pak 40, but the gun soldiered on till the end of the war.

Calibre:	50mm (1.97in)	Muzzle Velocity:	1198 m/s (3930 ft/sec)
Traverse:	65°	Armour Penetration:	A.P. 61mm at 500 yards (30°)
Elevation:	-8° to +27°	Armour Penetration:	A.P.40 86mm at 500 yards (30°)
Shell Weight:	A.P. 2.25kg (4lb 15¼ oz)	Complete Weight:	986kg (2,174lb)
Shell Weight:	A.P.40 0.97kg (2lb 2¼ oz)		

7.5cm Pak 40 (7.5cm Panzerabwehrkanone 40)
From November 1941 the standard anti-tank gun of the German ground forces

With the appearance of the Soviet KV-1 and T-34 tanks, a heavy, high-velocity weapon was needed. Already in its early design stage, development of the 7.5cm Pak 40 was given priority, and the first weapons were delivered in late 1941. It served on all

Rate of Fire:	14 rpm	Shell:	Armour-piercing/HE shell
Muzzle velocity:	792 m/s (2,598ft/s)	Calibre:	75mm (2.95in)
Weight:	1,425 kg (3,140lbs.)	Elevation:	-5° to +22°
Range:	1,800m (5,906 yards)	Traverse:	65°

7.5cm Pak 40 (7.5cm Panzerabwehrkanone 40)
From November 1941 the standard anti-tank gun of the German ground forces

fronts, and became the Wehrmacht's main anti-tank gun for the rest of the war. Some 23,500 were built, many being given to Germany's allies. An extra 6,000 were used to arm tank destroyers, and a lighter version (BK 7.5) was used in the Henschel Hs 129B-3 ground attack aircraft.

Number built: 23,500 approx
Length: 6.2m (20ft 4ins)
Range: 1,800m (5,906ft) direct fire
7,678m (25,190ft) indirect HE fire

8.8cm PAK 43 (Panzerabwehrkanone 43)
The most powerful anti-tank gun of World War II

With a virtually flat trajectory and very high velocity, the 8.8cm PAK 43 proved very easy to aim and achieve a first hit without the need to work out ranges. This gun out-ranged all Allied tank guns of the period. It could take out even the heavily-armoured Soviet IS-1 and IS-2 series tank and tank destroyers of the latter part of the war, at ranges of up to 2,500 yards, far outranging the Soviet weapons. The very low silhouette meant that once dug-in, it was very difficult to see. It could also be fired from its travel mount. Demand soon outstripped production, so in an effort to ease the situation it was mounted on the two-

Weight:	4,380kg (9,700lb)	Manufacturer:	Krupp, Rheinmetall
Length:	6.4m (21ft)	Range:	4,000m (4,400yds)
Barrel length:	6.61m (21ft 8in)	Rate of fire:	20-25 rpm
Height:	1.8m (5ft 11in)	Elevation:	-5° to +38°

PART VIII - ARTILLERY AND CREWED ANTI-TANK WEAPONS

8.8cm PAK 43 (Panzerabwehrkanone 43)
The most powerful anti-tank gun of World War II

wheel split-trail carriage from the 10.5cm leFH 18 howitzer. This version (Pak 43/41) proved heavy and difficult to manoeuvre in the mud and snow of the Eastern Front. The weapon was also mounted on a number of different tank and tank destroyers: in this format it was known as the KwK 43 L/71. Using the improved 8.8cm round, the armour penetration at long range was very impressive. The gun proved to be both liked by its crews and feared by its opponents, earning its reputation as the finest anti-tank weapon of the conflict.

Ammunition:	Pzgr. 40/43 APCR		Ammunition:	Pzgr 39/43 APCBC-HE
AP	Composite rigid construction		AP:	AP capped with ballistic cap - HE
Projectile weight:	7.3kg (16lb)		Projectile weight:	10,4kg (22.92lb)
Muzzle velocity:	1,130m/s	(3.707ft/s)	Muzzle velocity:	1,000m/s (3.281ft/s)

Rheinmetall 12.8cm K44 L55
Heavy anti-tank gun

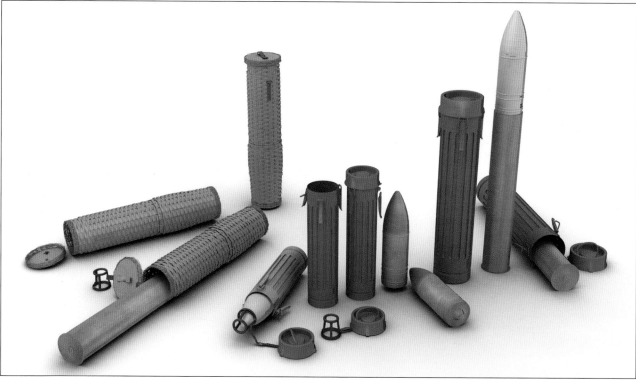

In 1943, Krupp was given a contract to supply a new heavy anti-tank gun, using tooling already available from their work on naval weaponry. The 12.8cm Panzerabwehrkanone came about as a direct result of reports from the Eastern Front, where for the first time the Germans found themselves outranged and at times outgunned by superior Soviet tank weaponry. With the introduction of the IS-2 heavy tank a gun was required that could outrange all Soviet tank guns. Using two-piece ammunition,

Rate of fire:	2-3 rpm	Shell:	Case separated loading
Muzzle velocity:	935 m/s (3,070 ft/s)	Calibre:	12.8mm (5in)
Weight:	210,160kg (22,400lb)		
Range:	24,410m (26,700yds)		

Rheinmetall 12.8cm K44 L55
Heavy anti-tank gun

Sitting on its cruciform platform, its very low silhouette of 4ft 7inches makes it an ideal anti-tank weapon: hard-hitting and difficult to see.

this gun in an artillery role could fire APCBC-HE shells, using light or medium charges, and PzGr.43 armour-piercing rounds using the heavy charge. At 1,000m, it could penetrate 200mm of 30° sloped armour and at 2,000m, 150mm of 30° sloped armour. The gun became the base from which many future tanks and tank destroyers would be designed. Like the Maus and E-100, though, the war end would before any of the derivations could go into combat.

Traverse:	360°	Designed:	1943
Elevation:	-7° 51' to +45° 27'	Manufacturer:	Krupp
Recoil:	Hydropneumatic		
Number built:	51		

2.8cm Schwere Panzerbüchse 41 (2.8 sPzB 41)
Light anti-tank gun using the squeeze bore principle

Standard issue: Officially classified as a heavy anti-tank rifle (Schwere Panzerbüchse) while really being a light anti-tank gun.

Version Two: without shield and small wheels extra light weight version for Fallschirmjäger units

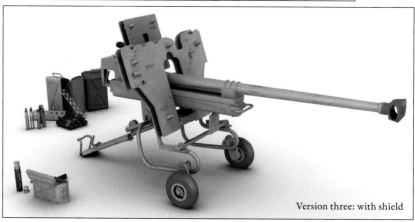

Version three: with shield

This gun was in service till late 1944, but then the lack of tungsten, used as the main core in its AT shells, caused production to stop. Some 2,797 were manufactured between 1933 and 1944. This weapon was also mounted on several vehicle types, such

Rate of fire:	25/30 rpm	Shell:	Armour-piercing/fragmentation
Muzzle velocity:	1,400m/s (4,593ft/s)	Calibre:	28mm (1.10in)
Weight:	1,800kg (4,000lb)		
Range:	500m (547 yards)		

PART VIII - ARTILLERY AND CREWED ANTI-TANK WEAPONS

2.8cm Schwere Panzerbüchse 41 (2.8 sPzB 41)
Light anti-tank gun using the squeeze bore principle

Late version 1943

sPzB 41 - Armour piecing shells showing Tungsten shot and the Sprg 41 fragmentation shell

Transit cases and packaging tubes

as half-tracks, armoured cars and even on off-road cars such as the Horch 901 4x4. The shell was fired using the squeeze bore principle. Its two external flanges were compressed by the tapered barrel, reducing the shell diameter and maintaining a high muzzle velocity.

Rate of fire:	25/30 rpm	Shell:	Armour-piercing/fragmentation
Muzzle velocity:	1,400m/s (4,593ft/s)	Calibre:	28mm (1.10in)
Weight:	1,800kg (4,000lb)		
Range:	500m (547 yards)		

91

7.5cm LG 40 Recoilless Gun (Airborne)
Short rifled howitzer used by airborne troops

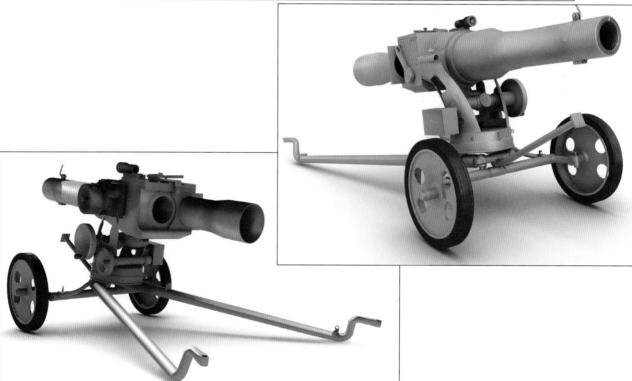

The short rifled howitzer had no recoil or counter-recoil system, as gases were vented to the rear through a funnelled tube, thereby equalising the pressure and eliminating any recoil. It was of a tubular steel construction, with weight kept to a minimum by the use of hollowed-out machine parts and plastic washers, with an aluminium alloy body. The wheels could be removed for air transport. This weapon was frequently short of ammunition, and reliant on the next supply drop.

Weight:	Travel 325lb	Muzzle velocity:	1,197ft/sec (HE)
Weight:	Firing 321lb	Elevation:	-15 +42° max
Weight of shell:	HE (12lb 9oz)	Traverse:	30° x 2 - 360°
	APC (15lb)	Horizontal range:	7,400yds
	Hollow charge (10lb 2oz)		

PART VIII - ARTILLERY AND CREWED ANTI-TANK WEAPONS

8.8cm Raketenwerfer 43 'Püppchen'
Light-weight field-piece for firing the 8.8cm Panzerschreck rocket

This field piece was designed and produced to increase the range of its 8.8cm hollow charge rocket. Using a percussion fuse, it fired a modified rocket up to a range of 700 yards. In action the wheels could be removed and the weapon rested on its rocker sledges. During winter the wheels could be replaced with skis for movement across snow-covered terrain. The rocket was loaded into a breech like a conventional shell and fired with very little recoil. It was much liked by German airborne forces, but only issued in small numbers because of the cost.

Weight:	149kg (325lb)	Weight of charge:	2.6kg (5lb 13oz) (TNT)
Length overall:	2.87m (8ft 10in)	Max range:	230m (700 yds)
Traverse:	60°	Armour penetration:	At 30° 250mm +/-
Elevation:	-18° to 15°	Introduced:	1943

7.5cm Leichte Infanteriegeschütz 18
Infantry support gun

This weapon, designed for mountain troops who had two guns per battalion, was capable of being broken down into six to ten packs and then carried by horse or mule. Some 12,000 were built between 1932 and 1945. An airborne version was also produced which broke down into loads of 4x 140kg. An infantry support gun (L/13) was also manufactured and tested but

Rate of fire:	8-12rpm	Shell:	Cased cartridge 6kg (13.2lb)
Muzzle velocity:	210m/s (689ft/s)	Calibre:	75mm (3.0in)
Weight:	440kg (970lb)		
Range:	3,550m (3,882yds)		

7.5cm Leichte Infanteriegeschütz 18
Infantry support gun

never adopted. Designed in 1927 by Rheinmetall, this weapon went on to serve on all fronts through the early years of the war. It was unique in having the barrel pivot at one end to open the breech mechanism for loading the shell.

Rate of fire:	8-12rpm	Shell:	Cased cartridge 6kg (13.2lb)
Muzzle velocity:	210m/s (689ft/s)	Calibre:	75mm (3.0in)
Weight:	440kg (970lb)		
Range:	3,550m (3,882yds)		

7.5cm Gebirgsgeschütz 36 (7.5cm GebG 36)
Mountain howitzer gun

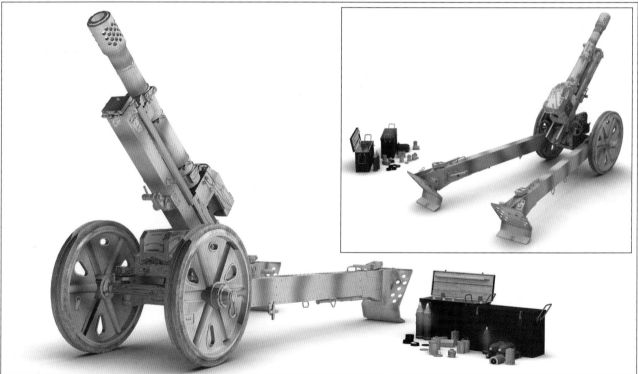

Another howitzer, designed by Rheinmetall to an army specification. This 7.5cm mountain gun used light alloys to reduce its weight: so much so that the recoil would make it jump excessively. At high angle firing the recoil was much less, as most of it was absorbed by the ground. It could be broken down into eight packs for mule transport. A mountain artillery battalion would have two or three batteries with four guns each. A regiment would consist of one to three battalions. This gun was used on all fronts.

Rate of fire:	6-8 rpm	Calibre:	75cm (3in)
Muzzle velocity:	475m/s (1,560 ft/s)	Crew:	Five
Weight:	750kg (1,700lb)		
Range:	9,250m (10,120yds)		

PART VIII - ARTILLERY AND CREWED ANTI-TANK WEAPONS

7.5cm Gebirgsgeschütz 36 (7.5cm GebG 36)
Mountain howitzer gun

Brass cartridges (early war years)
Steel cartridges (late war years)
Also shown, the anti-flash powder bag that was used to hid the guns position.

During the late war years, the ability to fire and manoeuvre so as to avoid counter-battery fire led to many field modifications. Its split tails could be shortened to make it fit the vehicle chassis.

The gun itself was a respectable weapon during the early war years, but its small 7.5cm calibre proved ineffective for the heavy demands of the late war years, where bigger guns, with a greater range and hitting power, were considered an operational necessity.

Shell:	75x130mm R (separate-loading, cased charge)	Number built:	1,193
Shell weight:	5.75kg (12.7lb)		
Elevation:	-2° to +70°		
Traverse:	40°		

10.5cm leFH 18 (leichte Feldhaubitze)
The standard light field howitzer gun

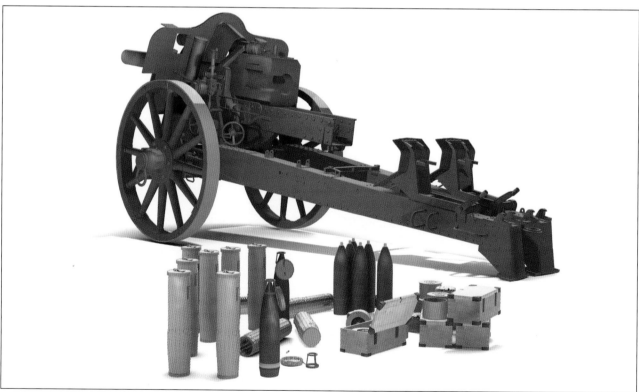

Designed and developed by Rheinmetall in 1930, this weapon went into service with the Wehrmacht in 1935 and went on to serve on all fronts during World War II. Indeed, many served in European armies into the 1980s. It was a solid, dependable design that was upgraded throughout its career while still retaining its basic core specifications. The original concept was for

Weight:	1,985kg (4,377lb)	Traverse:	56°
Caliber:	10.5cm (4.13in)	Rate of fire:	4-6rpm
Elevation:	-6° 30' to +40° 30'		

10.5cm leFH 18 (leichte Feldhaubitze)
The standard light field howitzer gun

a horse-drawn gun and limber, with wooden wheels. Later models used steel-pressed wheels, more suited for motor transport. It needed a crew of five.

Muzzle velocity:	470m/s (1,542ft/s)	Shell weight:	14.81kg (32.7lbs) HE TNT
Range:	10,675m (11,675yds)		14.25kg (31.4lb) AP
Shell:	Case separate-loading (6 charges)		

10.5cm Gebirgshaubitze 40 (10.5cm GebH 40)
Mountain howitzer gun

Designed by Böhler, this gun went into production during 1942 and went on to serve with the army's mountain divisions and the Waffen-SS. It saw action in the Balkans, the Eastern Front, Finland, Italy and the Western Front. It could be broken down into five loads for mule transport. At 1,660kg (3,700lb) it was heavy – the heaviest mountain howitzer ever to enter service – but it was regarded as one of the best mountain guns ever built, and many went on to serve in other armies until the mid 1960s. Some 420 were built between 1942-45.

Rate of fire:	4-6rpm	Range:	12,625m (13,807yds)
Muzzle velocity:	565m/s (1,850ft/s)	Shell:	Separate-loading, cased charge
Weight:	1,660kg (3,700lb)		

10.5cm Gebirgshaubitze 40 (10.5cm GebH 40)
Mountain howitzer gun

The carriage design was unique, in that the wheels were fixed to the split-tail arms. This made the wheels turn inward when the arms were opened and the gun was prepared to fire. This reduced the amount of backward movement when the gun recoiled as it was fired: a valuable feature as space on a mountainside is always at a premium. It fired three standard types of shell, 10.5cm hollow-charge armour piercing or high explosive shells and an illumination (star) shell. When firing high explosive, the propellant was added in six increments to reach the required range.

Calibre:	10.5mm (4.1in)	Barrel length:	2.87m (9ft 5in)
Shell weight:	14.52kg (32lb)	Number built:	420
Elevation:	-4°30' to +71°		
Traverse:	51°		

15cm sFH 18 (schwere Feldhaubitze 18)
Heavy field howitzer

Some 5,400 of the basic division-level heavy howitzer were produced. They were used thoughout the war on all fronts, and were held in high regard till the Eastern Front opened up. The gun proved to be inferior to the bigger 122mm and 152mm Soviet guns with their longer range: in some cases there was as much as a seven-kilometre range advantage.

Rate of fire:	4 rpm	Range:	13,250m (14,490yds)
Muzzle velocity:	495m/s (1,620 ft/s)	Shell:	Cased separate-loading ammunition
Weight:	5,530kg (12,191lb)		

15cm sFH 18 (schwere Feldhaubitze 18)
Heavy field howitzer

This was somewhat offset by the introduction of the rocket-assisted round (15cm R.Gr.19 FES) which went some way towards giving it parity with its Soviet counterparts.

Calibre:	149.1mm (5.89in)	Elevation:	-3° +45°
Barrel length:	4.5m (14ft 9in) L/30	Traverse:	64°
Recoil:	Hydropneumatic		

15cm sIG 33 (Schweres Infanterie Geschütz 33)
Standard heavy infantry gun of the German army

This was the largest gun ever to be classified as an infantry weapon by any nation. Originally manufactured with wooden wheels and horse-drawn, it was then modified to run on solid tyres with air brakes for towing behind motor vehicles. It was a workhorse of the German army and was used on all fronts up to the end of the war. Around 4,600 were built between 1936 and 1945.

A model was produced using lighter alloys - this reduced its heavy weight by 150kg (330lb) but was never adopted, as the Luftwaffe had priority on all light alloys. It needed a crew of 5-7.

Rate of fire:	2-3 rpm		Number built:	4,600 approx
Range of fire:	4,700m (5,100yds)		Muzzle velocity:	240m/s (790ft/s) HE
Weight:	1,800kg (4,000lb)		Shell:	1Gr 33 HE
Barrel length:	1.65m (5ft 5ins) L/11		Weight:	38kg (84lb)

15cm sIG 33 (Schweres Infanterie Geschütz 33)
Standard heavy infantry gun of the German army

The Stielgranate 42 was not a shell in the conventional sense: the round was loaded by inserting the driving rod into the gun barrel. The stabilising fins of the projectile remained outside the muzzle. A special charge – a blank round – was then loaded into the breech. When fired the driving rod would separate from the main body (around 150m, 160yds) leaving the projectile to fly to its target, up to a range of 1,000m (1,100yds). It was designed to demolish strongpoints and help clear minefields using its blast effect.

Filler:	8.3kg (18lb) amatol	Shell:	Stielgranate 42
Shell:	1Gr 38 hollow charge	Weight:	90kg (200lb)
Weight:	25.5kg (56lb)	Filler:	27kg (60lb) amatol
Filler:	Cyclonite/TNT		

21cm Mörser 18
Heavy howitzer

This huge weapon was transported in two pieces, the carriage and the barrel. Over small distances the whole gun would often be moved as a single entity. The dual recoil system was unique. As usual, the barrel recoiled in its cradle, but at the same time the whole top carriage, including the barrel, recoiled across the main carriage. This damped out the excess recoil.

When being set up to fire, the gun was lowered to the ground, to rest on its large round turntable while the wheels

Weight:	16,700kg (36,817lb)	Manufacturer:	Krupp
Length:	6.514m (21ft 4in)	Range:	14,500m (15,857yds)
Barrel length:	6.61m (21ft 8in)	Muzzle velocity:	550m/s (1,804ft/s)
Number built:	711+	Elevation:	-6° to +70°

21cm Mörser 18
Heavy howitzer

were cranked up off the ground. If the barrel had travelled separately, it was at this point that it was reinserted into the gun. Production stopped in 1942 in favour of the gun's smaller brother, the 17cm Kanone 18 Mörserlafette, which also had a range twice that of the 21cm Mörser 18, even with rocket-assisted shells. Production was resumed in 1943, more as a counter to Soviet heavy artillery, much of which outranged anything the Germans had.

Traverse:	On wheels, 16° On platform, 360°	Shell weight: Calibre:	113kg (250lb) HE 211mm (8.30in)
Shell:	Cased separate-loading ammunition (6 charges)		

15cm Nebelwerfer
Six-barrelled rocket launcher

Design and testing began in the 1930s, and this weapon entered service in 1940, in time for the battle for France. It was organised into batteries of six units to a company and three companies to a regiment. This weapon, with increases in range, calibre and warhead, would be used on all the main battlefronts. Over 5½ million rockets were manufactured during the conflict and some six thousand weapons built during the war.

HE:	15cm Wgr 41 Spr	Max range:	6,900m (7,500yds)
BCW:	15cm Wgr 41 Grünring	Introduced:	1940
Content:	chemical		
C.smoke:	15cm Wgr 41 Nb		

28cm/32cm Wurfkörper-Spreng
Artillery rocket, high explosive and napalm

28cm (High Explosive) and 32cm (Napalm)

A simple wooden 'firing' frame

This rocket came as a self-contained unit: the shell and the means to fire it. It had a wooden box frame which could be carried, with donkey legs that opened to give the required degree (5°-45°). It was fired by a hand-held electrical firing system. The safety pin was removed before firing and the flight fuse was a simple Wgr.Z. 50. The rocket jets at the base were angled at 12° from

28cm Wurfkörper-Spreng		Length overall:	3ft 11in
Shell:	110 lbs. (TNT)	Diameter (body):	11in
Range:	2,400 yds approx. at 42°		
Weight:	181lb		

WORLD WAR II GERMAN FIELD WEAPONS AND EQUIPMENT

28cm/32cm Wurfkörper-Spreng
Artillery rocket, high explosive and napalm

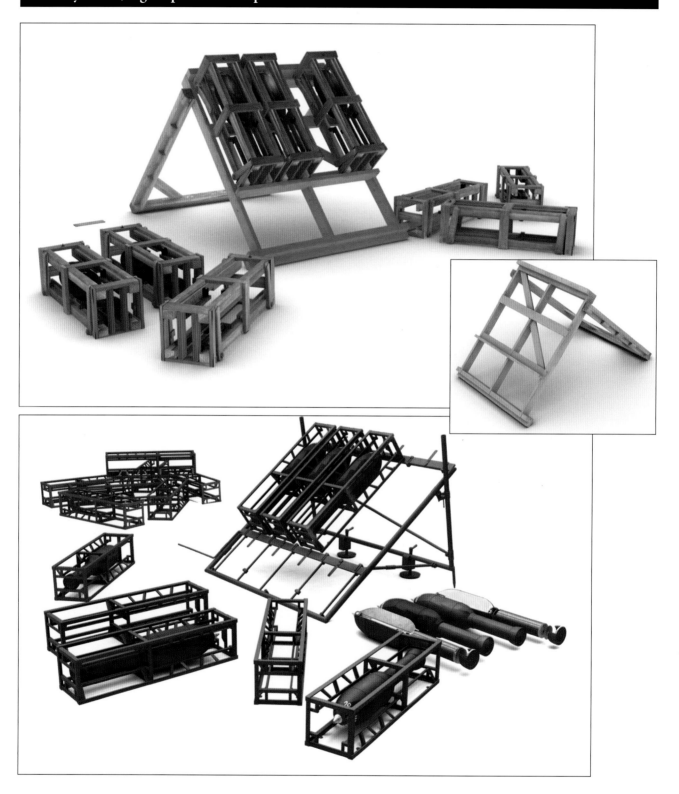

the diameter centre, giving the rocket rotation and stability in flight. This weapon was also fitted to a number of half-tracks (three being carried on each side) to improve mobility and the survival rates of their crews.

32cm Wurfkörper-Spreng
Shell: Napalm (incendiary)
Range: 2,400 yds approx. at 42°
Weight: 181lb

Length overall: 32ft 11in
Diameter (body): 11in Text

PART VIII - ARTILLERY AND CREWED ANTI-TANK WEAPONS

30cm Raketenwerfer 56
Mobile rocket launcher for 30cm rockets

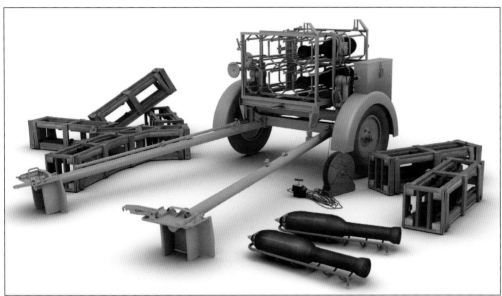

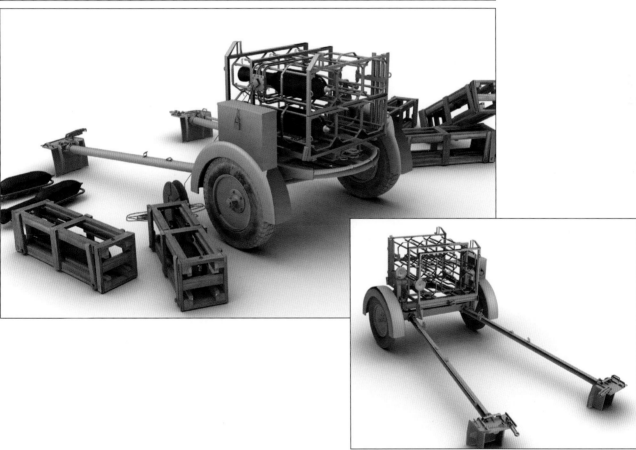

This was a six-barrelled rocket launcher mounted on the carriage of the 5cm PaK 38 anti-tank gun. The 30cm Wurfkörper 42 Spreng (explosive missile) was a spin-stabilised rocket: these were fired singly in a timed ripple by electrical firing. Its distinctive sound and huge smoke trail meant that when the last rocket was fired, there was a hurried move to avoid counter-battery fire. Some 694 were built from 1944 to 1945 and were used in almost all theatres.

Calibre:	301mm (11.9in)	Range:	4,550m (4,980yds)
Barrels:	6	Filling:	HE, 45 kg (99lb)
Elevation:	-3° to +45°		
Traverse:	22°		

Part IX - Anti-aircraft guns
Flak guns for mobile warfare

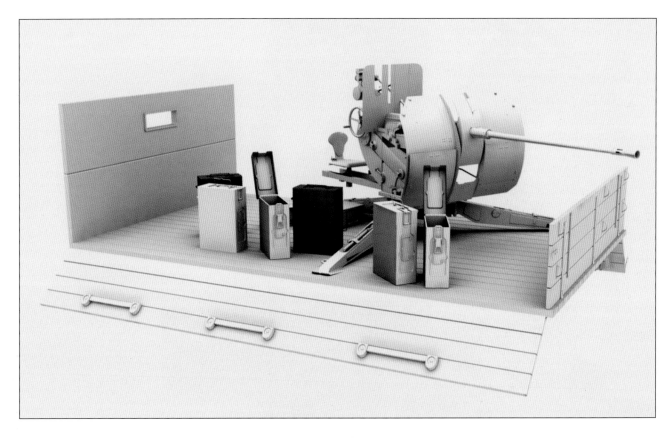

From the introduction of the internal combustion engine to the onslaught of World War I, motorised transport has been used to moved men and guns into battle. From the London bus ferrying troops to the Somme front, via machine guns in motorcycle side-cars roaming the Belgian countryside, to the first anti-aircraft guns being placed on the back of any available lorry or heavy tractor, improvisation occurred in response to local battlefield operations.

In World War II, this improvisation was taken to new heights and often resulted in official adoption of an idea, which was then put into standard practice.

The use of light anti-aircraft guns on the back of trucks soon became standard throughout combat zones. Today we see the same practical field improvisation in the use of heavy calibre machine guns and light anti-aircraft guns on the back of pick-ups in all modern conflicts from Africa to the Middle East.

PART IX - ANTI-AIRCRAFT GUNS

2cm Flak 38
20mm cannon with its two-wheeled trailer, the Sonder-Anhänger 51

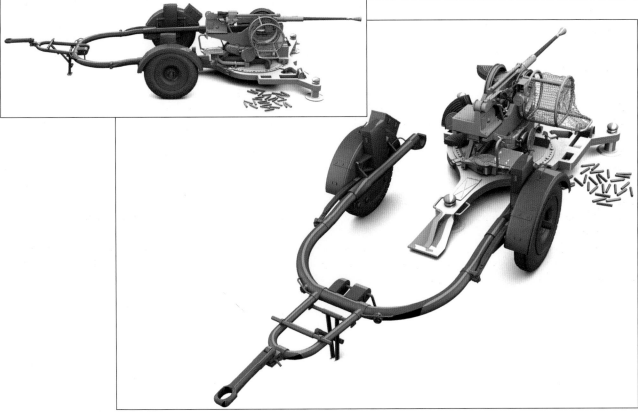

The 2cm Flak 38 was a light anti-aircraft cannon, standard issue to all infantry divisions at the beginning of the war. Each division was issued with 84 such weapons. Anti-aircraft regiments had 60 guns allocated to them. This weapon could also be used against ground targets.

Weight:	483kg (10lb 65oz)	Elevation:	90° max
Weight of shell:	120gm (4lb 2oz)	Horizontal range:	2,697m (2,950 yds)
Rate of fire:	1,800 rpm	Vertical range:	3,214m (7,000 yds)
Muzzle velocity:	899m/sec (2,950ft/sec)		

2cm Flakvierling
20mm quad cannon

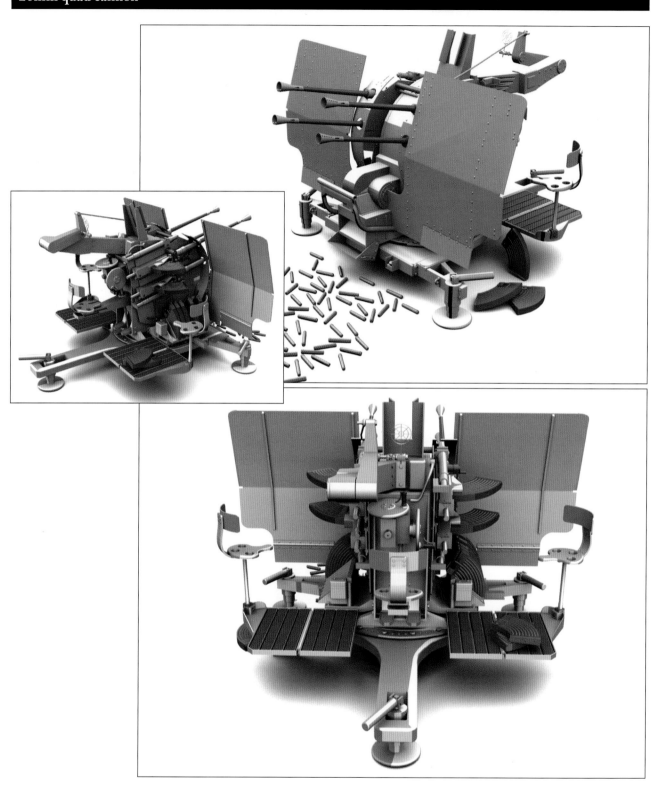

The 2cm Flak was developed by the Mauser company. This four-barrelled flak cannon was adopted by all arms of the German armed forces. A large number were also fitted to half-track and flat-bed trucks to increase the protection of mobile columns. A number were also fitted to trains, to provide cover against air attack.

Weight:	1,521kg (3,352lb)	Elevation:	100° max
Weight of shell:	120gm (4lb 2oz)	Horizontal range:	2,697m (2,950yds)
Rate of fire:	1,800 rpm	Vertical range:	3,214m (7,000yds)
Muzzle velocity:	899m/sec (2,950ft/sec)		

PART IX - ANTI-AIRCRAFT GUNS

Schwerer Wehrmachtschlepper (s.W.S.)
Late production half-track with quad 2cm Flak

Designed as a replacement for all half-track series of vehicle and to standardise all half-track production, this robust vehicle had many variants, including carrying a 15cm Nebelwerfer rocket launcher, a 3.7cm Flak or, as displayed here, the 2cm Flakverling.

Weight:	9.5 tons	Width:	250cm
Crew:	From 2-6	Speed:	28km/h road
Length:	667cm	Fuel load:	240 litres
Height:	285cm		

2cm Gebirgs-Flak 30
Special-purpose 20mm anti-aircraft cannon

At the beginning of the war, the 2cm Flak 30, a light anti-aircraft cannon, was standard issue to all infantry divisions. The Gebirgs-Flak 38 model was issued to mountain and airborne units.

Weight:	360kg	Horizontal range:	2,200m (2,406 yds)
Rate of fire:	450 rpm		
Muzzle velocity:	900m/sec (2,950ft/sec)		
Elevation:	90° max		

3.7cm Flak 36 and 37

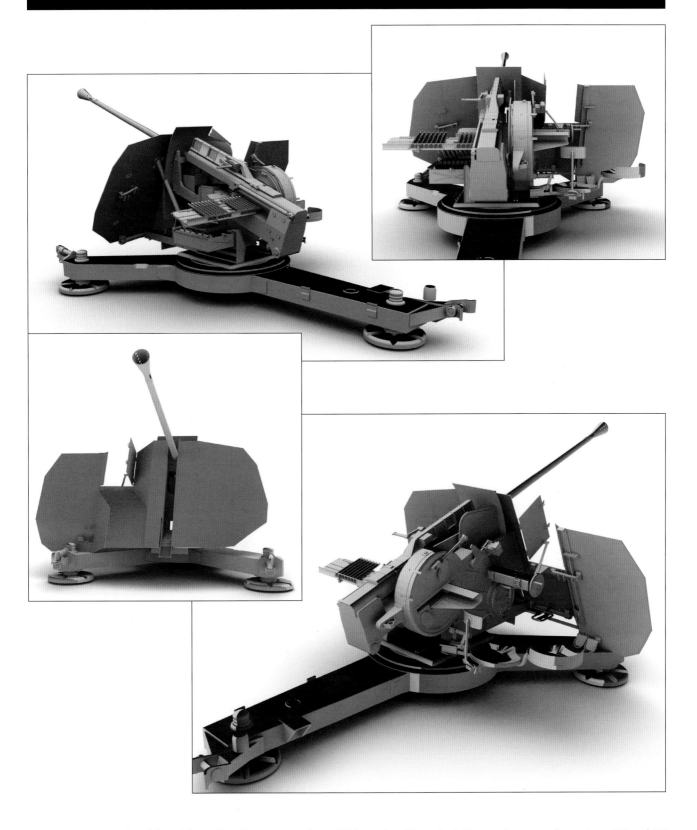

This was an upgrade of the Flak 18, based on reports from field combat. The sole difference between the variants 36 and 37 was in the type of gunsight fitted.

Weight:	1,544kg (3,405lb)	Elevation:	85° max
Weight of shell:	556gms (19lb 6oz)	Horizontal range:	6,492m (7,100 yds)
Rate of fire:	160 rpm	Vertical range:	4,785m (1,570 yds)
Muzzle velocity:	820m/sec (2,690ft/sec)		

Truck Platform for Light Flak
Wooden sledge for placement of Flak guns on flatbed trucks to increase mobility

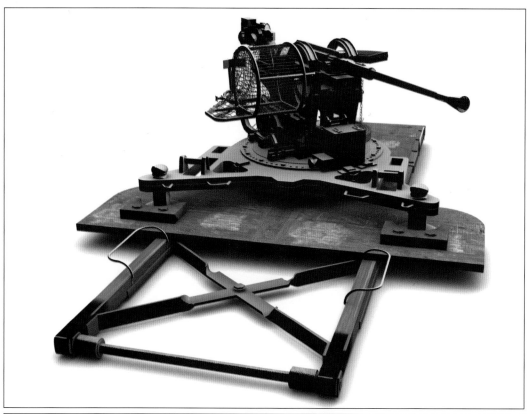

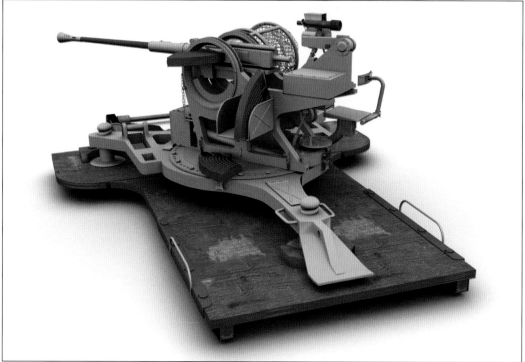

The wooden sledge was a field improvisation to reduce wear and tear on the flatbed trucks used as mobile gun platforms – a job for which they were not designed. It was soon adopted by all arms of the German forces.

5cm Flak 41

PART IX - ANTI-AIRCRAFT GUNS

This weapon, designed by Rheinmetall-Borsig in 1936 was an attempt to close the 'air' gap, a band approximately 1,500m (4,900ft) to 3,000m (9,800ft) which small calibre weapons and large calibre weapon could only cover with difficulty. This weapon was one of the least successful of German flak guns, shown by the simple fact of only 60 being produced, though this weapon would remain in use till the end of the war. Its crew of seven had to hand-feed the five-round clips into the gun, not easy as the recoil on this gun was excessive due in part to its underpowered ammunition. Added to this the dazzling muzzle 'flash' when the gun was fired could blind/distract the gun-aimer, even in daylight. Its slow traverse and rate of fire (only 130 rounds per minute) made it difficult to follow fast-moving targets, let alone destroy them.

Weight (travelling):	7.18 tons	Weight of Shell:	H.E. 4.8lbs
Weight (firing position):	4.30 tons		A.P. 4.87lbs
Ammunition:	H.E. 41/tracer, Incendiary/	Muzzle Velocity (H.E.):	2,756 f/s
	H.E. 41, A.P.C.B.C. 42	Rate of fire:	130rpm

5cm Flak 41 mit Sd.Ah 204
Towed Anti-aircraft gun

Effective ceiling:	10,000ft	Manufacturer:	Rheinmetall-Borsig
Traverse:	360°	Produced:	1936
Elevation:	90°	Total Produced:	60
Calibre:	5cm	Number of Crew:	7

PART IX - ANTI-AIRCRAFT GUNS

8.8cm Flak 18/36
The nemesis of Allied tank crews

Designed by Krupp in 1928 as a mobile AA gun, this gun was capable of being brought into action in two and a half minutes. The success of the weapon in the early war years in its improvised role as an anti-tank gun, led it to becoming, in later years,

Rate of fire:	15-20rpm	Calibre:	88mm (3.46ins)
Range:	14,810m (16,200yd) ground target	Elevation:	-3° to +85°
	11,900m (39,000ft) max ceiling	Traverse:	360°

8.8cm Flak 18/36
The nemesis of Allied tank crews

the bogeyman of all tank crews. Such was its success that it began to be developed as a tank gun. Eventually the Elefant, Jagdpanther, Tiger and King Tiger were all equipped with upgraded 'eighty-eights'.

Muzzle velocity:	820m/s (2,690ft/s)	Number built:	21,310 approx
Recoil:	Independent liquid and hydropneumatic	Weight:	7,407kg (16,325lb)
Carriage:	Sonder-Anhänger 202	Length:	5,791m (20ft)

12.8cm Flak 40 Zwilling
Twin mounted heavy anti-aircraft guns

PART IX - ANTI-AIRCRAFT GUNS

Rheinmetall Borsig designed and developed this weapon from 1936, and the first went into service in 1942. Set up for firing, it weighed nearly 12 tonnes. Its special low bulk trailer used to move the weapon proved impractical over rough ground. A few were placed on railcars to allow some degree of movement, but most were placed in fixed positions with concrete bases to hold the mounts. The anti-aircraft Zoo Tower in Berlin had four twin mounts. Hamburg, Vienna and other major cities also had their fixed gun sites.

Rate of fire:	20rpm	Range:	10,675m (35,025 yds)
Muzzle velocity:	880m/s (2,887ft/s)	Elevation:	-3° to +88°
Weight:	17,000kg (37,478lb)	Traverse:	360°

12.8cm Flak 40 Zwilling
Twin mounted heavy anti-aircraft guns

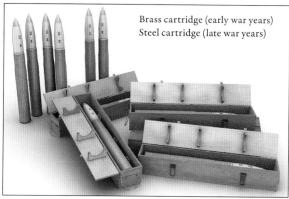

Brass cartridge (early war years)
Steel cartridge (late war years)

This 12.8cm gun fired a shell which used a powder charge four times greater than the 88mm Flak 18. This meant a shell flight time shortened by a third, thus making it easier to target fast-moving aircraft. This gun was used as artillery during the battle for Berlin in 1945.

Feed system:	Power rammer	Recoil:	Hydropneumatic
Calibre:	12.8cm (5.03in)	Shell weight:	27.9kg (52.2lb)
Length:	7.835m (25ft 8in)	Number built:	1,125

Part X – Vehicles
Supplying fuel to the armies of the Third Reich
'All modern armies are thirsty'

The world-famous jerrycan was copied by all combatants in World War II. Jerrycans with a white cross were used for water, all others carried petrol. Fuel drums with hand-cranked pumps were used for field refuelling.

Sd.Ah.S1
General purpose trailer

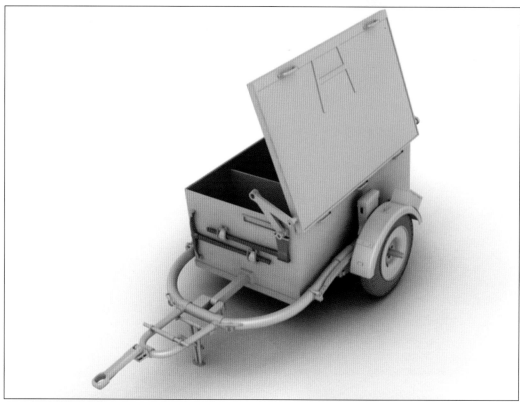

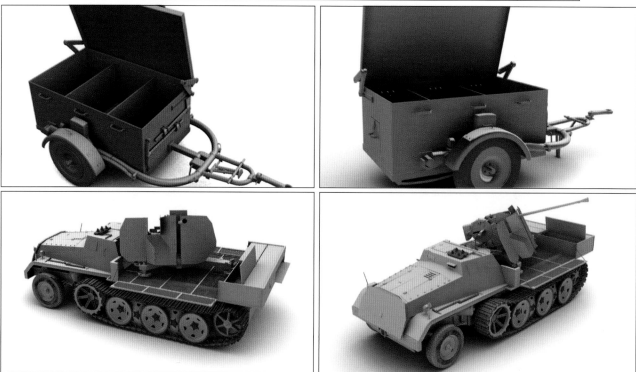

A general-purpose trailer, used as an extra stowage bin or for ammunition storage for mobile vehicle-mounted flak units, such as the 3.7cm Flak, mounted here on a late production model Schwerer Wehrmacht-Schlepper (s.S.W).

PART X – VEHICLES

Raupenschlepper Ost (RSO)
Fully tracked ambulance on an RSO chassis

After the first winter on the Russian front, it was clear that a vehicle was needed to deal with the harsh conditions: the muddy autumns, freezing, snowy winters and wet and muddy springs. Steyr introduced in 1942 the Raupenschlepper Ost (RSO), both as a prime mover and general supply vehicle. It was not long before purpose-built bodies were constructed to fit the existing chassis. The 7.5cm Pak 40/4 was mounted as a self-propelled gun, in addition to its use as a frontline ambulance suited to the difficult terrain and muddy conditions of the Russian Front.

Length:	4.25m	Engine:	Steyr V8 3.5l/8-cylinder petrol; 85hp
Width:	1.99m	Speed:	30km/h (18mph)
Height:	2.53m		
Crew:	2		

NSU Kettenkrad 'HK 101' (Sd. Kfz.2)
Light prime mover

The Kettenkrad was originally intended for towing light artillery guns, mortars and telephone wire drums. It was found useful by all branches of the German armed forces. A total of around 8,345 vehicles were produced during the war years.

Weight:	1280kg (1.3t)	Dimensions:	Length: 300cm
Crew:	1 to 3		Height: 100cm
Pulling capacity:	450kg		Width: 120cm
Speed:	70km/h		

PART X – VEHICLES

Bison - 15 cm sIG 33 auf Fehrgestell Panzerkampfwagen II
Self-propelled gun on a Panzer II Ausf B chassis

This was a modification from the early war years: the 15cm sIG 33 heavy infantry gun, married to the chassis of a Panzerkampwagen II Ausf B. It was first encountered during the invasion of the Low Counties in 1940. This weapon also saw extended service in North Africa, right up to the surrender of Axis forces in May 1943.

Weight:	11.2 tons	Manufacturer:	Alkett
Road speed:	40kph (25mph)	Main armament:	15cm sIG 33
Road range:	190km (120m)		
Number built:	12		

Renault UE Chenillette (selbstfahrlafette Pak 36)
Field improvisations of the commonest French armoured vehicle

After the fall of France, the German forces captured thousands of 'tankettes'. The commonest of these was the Renault UE Chenillette. On the principle of, 'to the victor the spoils', the captured French armaments began to supplement standard German issue. Over the next few years, some 700 conversions were made. The Selbstfahrlafette für Pak 36 auf Renault UE(f) was fitted with bucket-style clamps at the front for the gun's wheels. The crew serviced the gun relatively unprotected: ammunition was stored in the large bin at the rear.

Weight:	2.64 tons (5,800lb)	Shell:	Stielgranate 41
Road speed:	30km (19mph)	Shell range:	800m
Road range:	100km (62mi)		
Armament:	Pak 36		

PART X – VEHICLES

7.5cm Pak 40/4 auf Raupenschlepper (RSO)
Self-propelled anti-tank gun on tracked RSO chassis

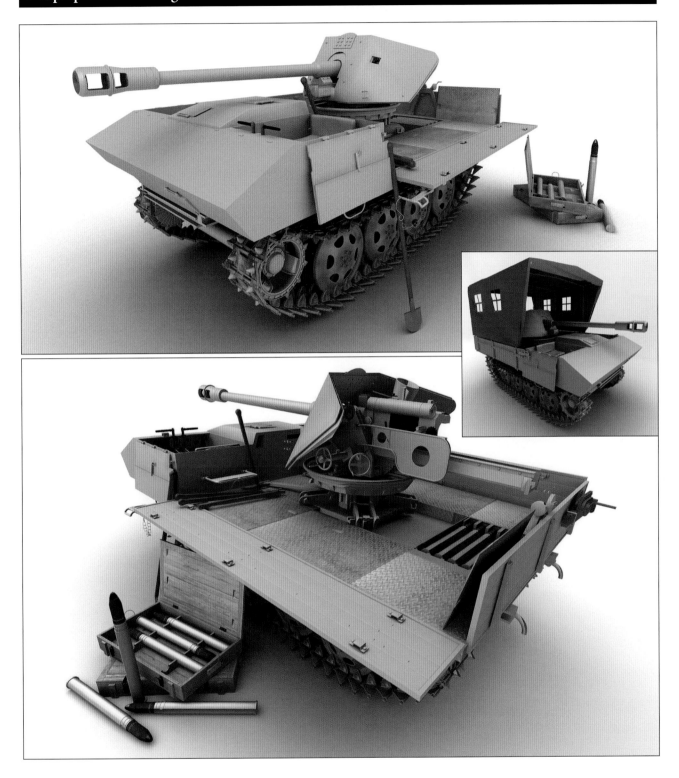

The normal RSO tracked lorry chassis was modified to take the Pak 40/4 anti-tank gun. The cab was replaced by a lightly-armoured version with an open driving position but drop-down armoured plates. While the vehicle gave reasonable cross-country performance, the standard Pak 40/4 could, even in 1944, be a lethal weapon to all Allied tanks at ranges up to 2,500 yards (up to 115mm armour penetration at 500 yards.)

Calibre:	7.5cm (2.95in)	Shell weight:	(AP40) 3.2kg (7lb)
Traverse:	65°	Introduced:	1941
Elevation:	-5° to +22°		
Range:	AP40 (115mm at 500 yards)		

Schwerer Wehrmacht-Schlepper (s.W.S)
Late production half-track with 15cm Nebelwerfer

This was a late production model of the Schwerer Wehrmacht-Schlepper (s.S.W.). It was armed with a standard 10-barrelled 15cm Nebelwerfer. With a crew of four, this proved to be a very effective weapon. The weapon carried ten rounds, with a further ten rounds inside the vehicle. The range of the rocket projectile was 73,000 yards and the HE round had a 28lb bursting charge.

Weight:	9.5 tons	Width:	250cm
Crew	From 2 to 6	Speed:	28km/h road
Length:	667cm	Fuel capacity:	240 litres
Height:	285cm		

PART X – VEHICLES

15cm Panzerwerfer 42 (Sf.) auf LKW Opel 'Maultier' (Sd.Kfz.4/1)
Ten-barrelled Nebelwerfer mounted onto a lightly armoured vehicle for mobility

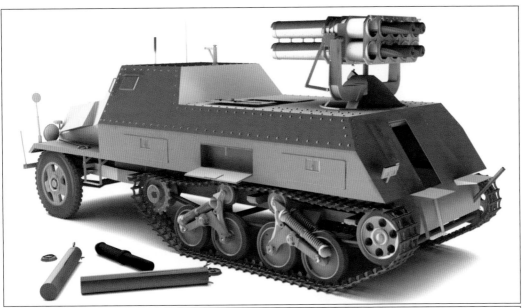

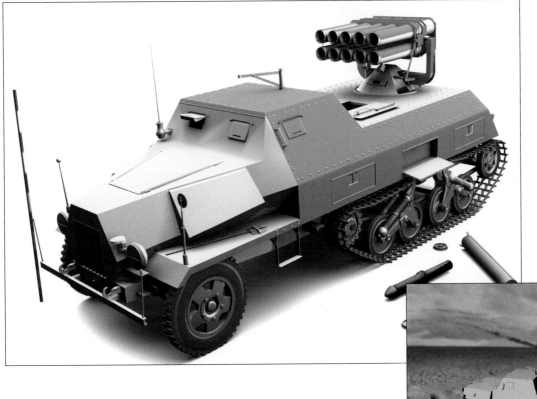

To improve both mobility, and the survival capacity of its Nebelwerfer units, 300 Opel-Blitz 'Maultier' chassis were converted into light armoured mobile rocket launchers. With a crew of four, this proved to be a very effective weapon. Ten rounds were carried in the weapon and a further ten rounds inside the vehicle.

Weight:	7.1 tons	Width:	7.52ft
Crew:	4	Speed:	24km/h road
Length:	19.98ft	Fuel load:	240 litres
Height:	8.95ft		

Goliath (Leichte Ladungsträger Sd.Kfz.302)
Guided remote-controlled tracked demolition charge

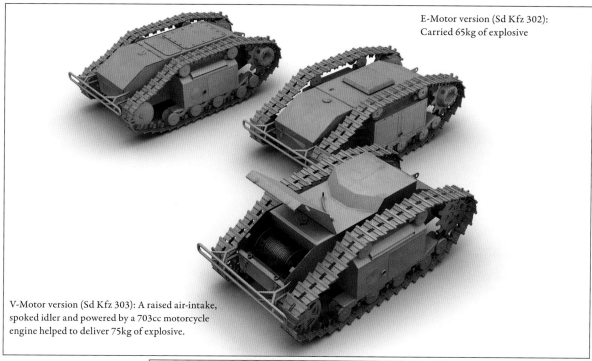

E-Motor version (Sd Kfz 302): Carried 65kg of explosive

V-Motor version (Sd Kfz 303): A raised air-intake, spoked idler and powered by a 703cc motorcycle engine helped to deliver 75kg of explosive.

Late war years control box and firing key

Early war years, control box

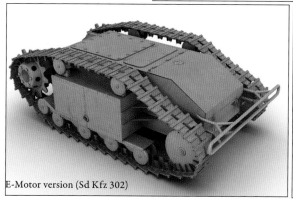

E-Motor version (Sd Kfz 302)

Showing the cable spool compartment

Goliath was an expendable guided remote-controlled bomb, first designed for clearing minefields and later used for general demolition of strongpoints, buildings, etc. Control was manual, via a small box. Of the three-strand wire (held in the cable compartment at the rear of the vehicle) two were used to steer, blue for the left and white for the right. The last wire (red) was used to detonate the 65/75KG charge. Some 2,650 were produced, from April 1942 to January 1944.

Weight:	0.37 tons	Speed:	10km/h road
Length:	1.5m	Range:	1.5km
Height:	0.56m	Engine:	Two Bosch MM/RQL electric 2.5kw
Width:	0.85m		

PART X – VEHICLES

Sd.Kfz.304 NSU 'Springer' Mittlere Ladungsträger
Expendable armoured demolition vehicle

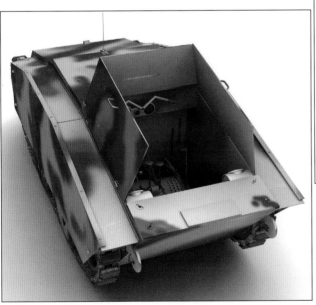

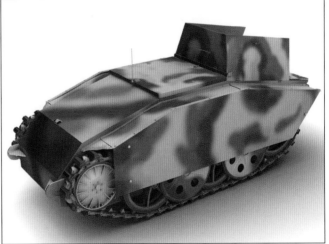

This expendable vehicle, which carried an explosive charge of 500kg, was designed to clear minefields and demolish fortifications and pillboxes. The driver placed the vehicle as close to the target as possible, left the vehicle and then guided the vehicle to the target by radio. The charge was then fired by a signal from the radio in the vehicle. Nearly 50 vehicles were produced, from October 1944 to February 1945. The Springer used many parts from the Sd.Kfz.2 Kettenkraftrad, including the drive-train and engine parts.

Weight:	2.4 tons	Speed:	42km/h road
Length:	3.17m	Range:	200km
Height:	1.45m	Engine:	Opel Olympia 1.5lit
Width:	1.43m	Radio:	KE6 mit UKE6

Borgward IV (Schwere Ladungsträger (Sd.Kfz.301 Ausf C)
Tracked demolition charge-layer

Introduced in 1943, this vehicle was designed to be simple to construct and maintain. It also used less armour-plate overall. It could deliver a main explosive charge of 500kg by the means of drive and drop. TV cameras were also trialed to see if they helped the driver improve his ability to deliver the explosive charge to the target while huddled down and under fire, but nothing came of this.

Weight:	4.85 tons	Speed:	40km/h road
Length:	4.1m	Range:	212km
Height:	1.25m	Engine:	Borgward 6B 3.8lit
Width:	1.83m	Radio:	EPS mit UKE6

'Wanse'
Self-propelled anti-tank vehicle

As the war reached Berlin, many of these tracked demolition-charge layers were converted into small self-propelled weapon platforms by the addition of a 8.8cm rocket launcher. Smoke discharges were fitted to the front of the chassis to help lay a smoke screen, behind which the vehicle could retire after ambushing and firing its salvo of rockets.

Being a small vehicle with a low profile and a reasonable turn of speed, it had some success as a tank hunter during the battle for Berlin in 1945.

Weapons:	8.8cm	Raketenpanzerbüchse	54/1	Armour penetration: At 30° 250mm +/-
(Panzerschreck)				Introduced: 1943
Weight of charge:	2.6kg (5lb 13oz) (TNT)			
Max range:	230m (700 yds)			

Kugelpanzer (Observation Tank)
Prototype of armoured tank for a forward artillery observer to use

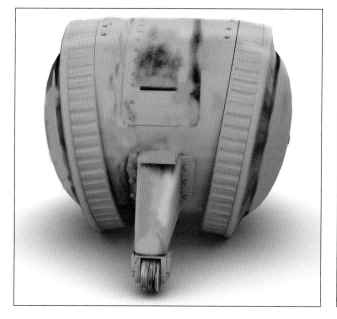

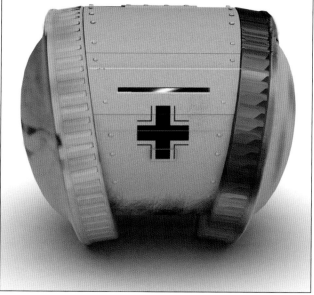

This vehicle never went beyond the prototype stage and was never issued to any units. It was powered by a two-stroke motorcycle engine and lightly armoured. The only known prototype is held at the Kubinka Tank Museum outside Moscow, Russia. The existence of this vehicle only became known in the 1990s when the Communist system ended and the Russia opened its hidden museums.

PART X – VEHICLES

Alkett VsKfz617 Minenräumer
1943 prototype of mine-clearing vehicle

An unknown prototype, seized by the Soviets for evaluation purposes, only became known to the West after the ending of Communism in the Soviet Union. Designed by Alkett, Mercedes-Benze and Krupp in 1942. its very excessive weight slow speed and high profile made it an unsuitable vehicle for combat. None were ever issued to frontline units. The only known prototype is held at the Kubinka Tank Museum outside Moscow, Russia.

Length:	6.8m (20,724 ft)	Armament:	7.92mm MG34 x 2
Width:	3.22m (10,626 ft)		
Height:	2.90m (9.57 ft)		
Armour:	10-40mm (up to 1.6 in)		

Part XI – Late production munitions and miscellaneous kit
Infrarot-Scheinwerfer 1945
Infra-red night-vision devices 1945

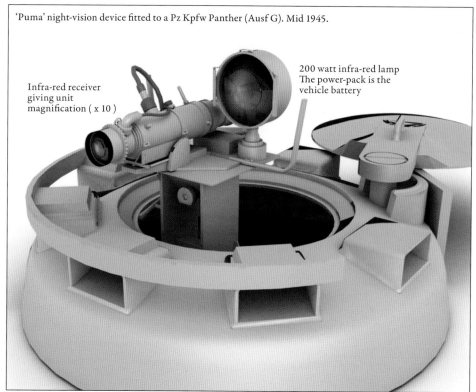

'Puma' night-vision device fitted to a Pz Kpfw Panther (Ausf G). Mid 1945.

Infra-red receiver giving unit magnification (x 10)

200 watt infra-red lamp
The power-pack is the vehicle battery

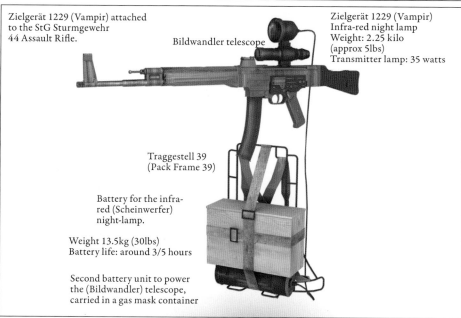

Zielgerät 1229 (Vampir) attached to the StG Sturmgewehr 44 Assault Rifle.

Bildwandler telescope

Zielgerät 1229 (Vampir)
Infra-red night lamp
Weight: 2.25 kilo (approx 5lbs)
Transmitter lamp: 35 watts

Traggestell 39 (Pack Frame 39)

Battery for the infra-red (Scheinwerfer) night-lamp.

Weight 13.5kg (30lbs)
Battery life: around 3/5 hours

Second battery unit to power the (Bildwandler) telescope, carried in a gas mask container

In late 1944, AEG, Letiz and RPF produced a night-vision device (infrared sights) for both AFVs and small arms. Only the Stg 44 Assault Rifle was fitted with what is a very heavy and awkward load to carry. The visual range of the scope was around 100 yards (clear image). The soldiers who used these were called Nachtjäger (night-hunters). The scope and lamp fitted to the tank (and other AFVs) gave an image up to 800 yards. There were many variations on the theme of infrared, but most never got past the drawing board.

Fliegerfaust
Late-war anti-aircraft weapon 1945

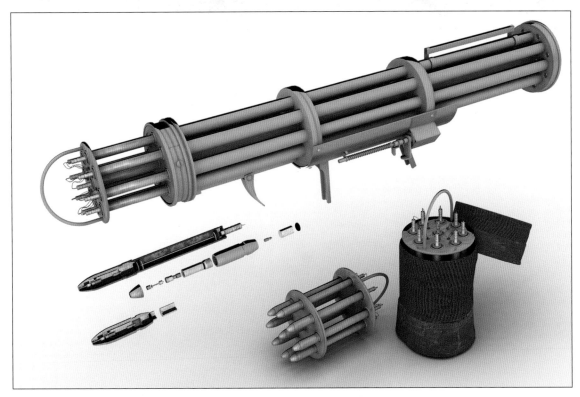

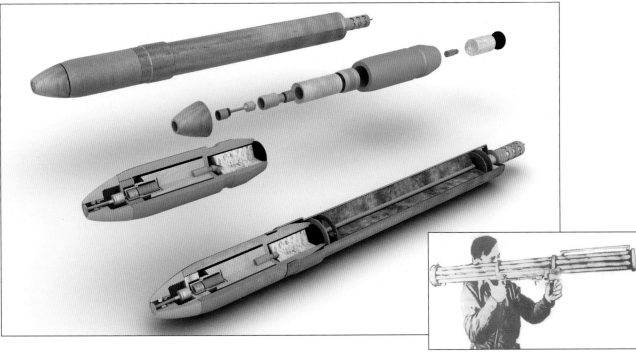

This weapon was made by the armaments company HASAG (Hugo Schneider AG) in 1944. Of the 10,000 launchers and four million rockets which were ordered, only 80 of these weapons were issued, and the only existing combat photograph shows one discarded in the rubble after the fall of Berlin in 1945. First, four rounds were simultaneously fired by every second barrel and after a time delay of 0.1 seconds, the remaining five rounds were launched. This was to avoid the rocket's exhaust fumes interfering with or damaging the rounds. It was not a successful weapon.

Shell: 20mm x nine rounds
Range: 500m
Weight: 90g
Small rocket length: 150cm

Total weapon weight: 65k

X7 Rotkäppchen
The world's first wire-guided anti-tank missile, 1945

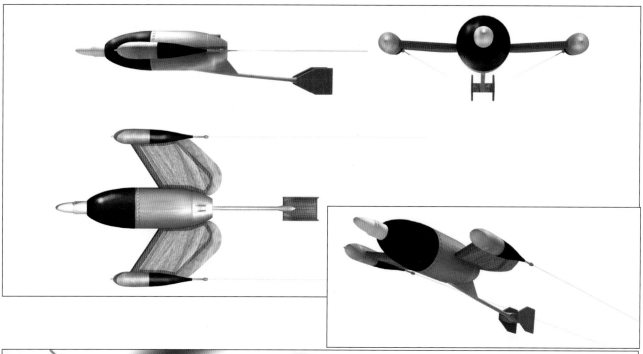

Under the leadership of Max Otto Krammer, (Ruhrstahl AG) the X-7 Rotkäppchen (Red Riding Hood) anti-tank missile was developed from 1943 onwards, for use against tanks and other armoured vehicles. The WASAG 109-506 solid fuel rocket was used to thrust the rocket to a range of two kilometres and its 2.5kg shaped charge warhead could defeat most Allied tanks of the period. A few hundred of these weapon were manufactured and there are unconfirmed reports that after flight-testing some were used with great success on the Eastern Front.

Weight:	9kg		Wingspan:	0.6m
Weight of charge:	2.5kg		Engine:	WAS AG 109-506 solid fuel
Diameter:	0.15m		Range:	2km
Length:	0.95m		Fuse:	Impact

PART XI – LATE PRODUCTION MUNITIONS AND MISCELLANEOUS KIT

Rheinbote Missile (Rhine Messenger)
Surface to surface artillery missile 1944

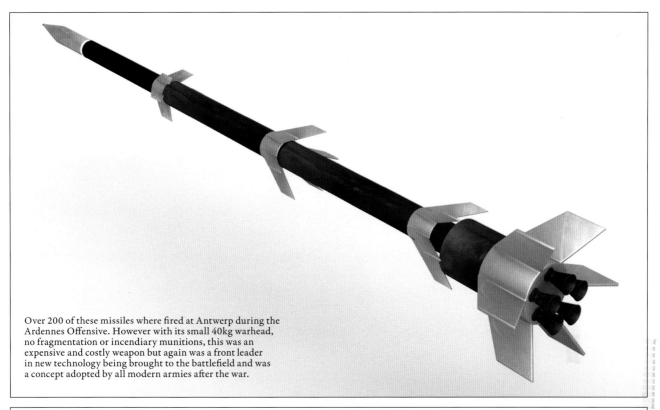

Over 200 of these missiles where fired at Antwerp during the Ardennes Offensive. However with its small 40kg warhead, no fragmentation or incendiary munitions, this was an expensive and costly weapon but again was a front leader in new technology being brought to the battlefield and was a concept adopted by all modern armies after the war.

Designed to be mobile and often fired from converted artillery platforms, the cruciform platform for the 88mm gun was one such platform used. Another was the Mobile firing platform used by the V.2.

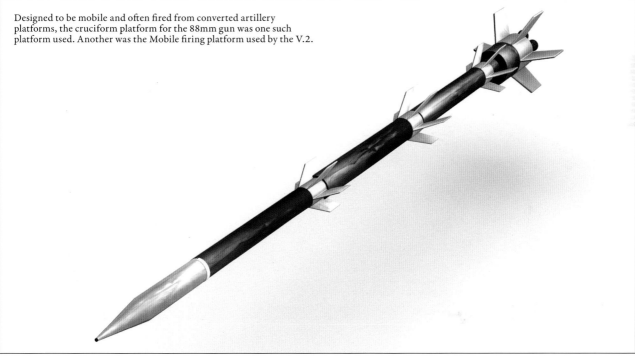

This four-stage rocket was developed by Rheinmetall-Borsig, and it first flew in late 1943. Its basic premise was to provide very long-range artillery fire for mobile troops hence its need to be mobile and lightweight. It had a range of 220km. Its speed of Mach 5.5 was not beaten until the introduction of ICBMs by the Soviets and Americans during the Cold War in the 1960s.

Launch weight:	1.709kg	Speed:	5.5mach
Weight of charge:	40kg	Rocket:	Diglycol solid propellant
Length:	11.4m		
Range:	220km		

7.3cm Propaganda-Werfer (Propaganda Rocket)

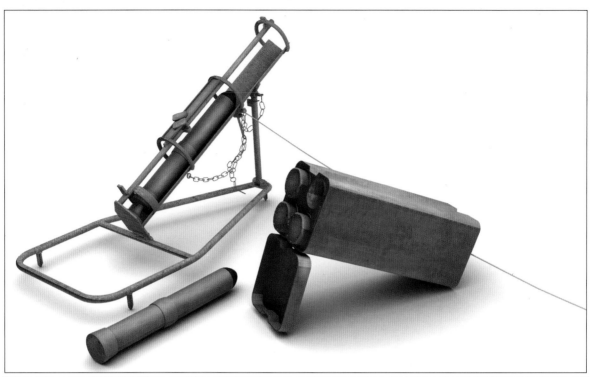

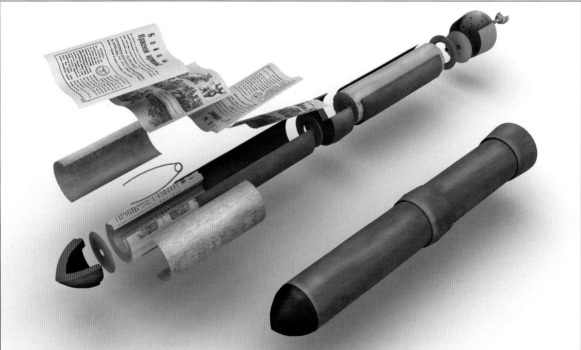

This was a small conventional rocket projectile, its warhead filled with propaganda leaflets. The rocket was placed on its launcher and hung (like a mortar round), till released by pulling its release lever, dropping the round onto its firing pin. The rocket consisted of two parts, the base, filled with the propellant, and the top half, containing 0.5kg of leaflets. These were wrapped around a steel spring so that when a small charge burst the top apart into two halves, the spring expelled the leaflets, spreading them out over a wide area.

Weight: 27lb
Weight of round: 6lb 10oz
Weight of propellant: 1lb
Overall length: 16 3/32in

Launcher weight: 27lb